FACTS AT YOUR FINGERTIPS

## THE WORLD OF ENDANGERED ANIMALS

# NORTH AND SOUTH AMERICA

**Published by Brown Bear Books Limited**

4877 N. Circulo Bujia
Tucson, AZ 85718
USA

and

First Floor
9-17 St. Albans Place
London N1 ONX
UK



Library of Congress Cataloging-in-Publication Data available upon request.



ISBN-13 978-1-78121-075-8

In this book you will see the following key at top left of each entry. The key shows the level of threat faced by each animal, as judged by the International Union for the Conservation of Nature (IUCN).

| | |
|---|---|
| **EX** | Extinct |
| **EW** | Extinct in the Wild |
| **CR** | Critically Endangered |
| **EN** | Endangered |
| **VU** | Vulnerable |
| **NT** | Near Threatened |
| **LC** | Least Concern |
| **O** | Other (this includes Data Deficient [DD] and Not Evaluated [NE]) |

For a few animals that have not been evaluated since 2001, the old status of Lower Risk still applies and this is shown by the letters **LR** on the key.

For more information on Categories of Threat, see pp. 54–57.

**Editorial Director:** Lindsey Lowe
**Editor:** Tim Harris
**Design Manager:** Keith Davis
**Designer:** Lynne Lennon
**Children's Publisher:** Anne O'Daly
**Production Director:** Alastair Gourlay

**Picture Credits**

**Cover Images**
**Front:** *Golden lion tamarin,* Eric Gevaert/Shutterstock
**Back:** *Galápagos Marine Iguana,* Paul van den Berg/Shutterstock

**Corbis:** Pete Oxford/Minden Pictures 44–45; **FLPA:** Kevin Schafer/ Minden Pictures 8–9; **Shutterstock:** Eric Gevaert 1, John A. Anderson 4–5, San Rafael Waterfall 7, Fotomicar 13, Eric Gevaert 15, Steve Oehlenschlager 18–19, Olga Vasik 20–21, Dennis Donohue 23, Jean-Edouard Rozey 24–25, Sergi Lopez Roig 27, Mogens Trolle 29, Buchan 30–31, Critterbiz 32–33, Stubblefield Photography 35, Worlds Wildlife Wonders 37, 39, Matt Jeppson 47, Photovolcania.com 48–49, Alfredo Maiquez 51, Galyna Andrushko 54–55, Chaos Maler 57, Jerry Sanchez 59.



Printed in the United States of America

062014
AG/5534

# CONTENTS

# Habitats, Threats, and Conservation

Every major kind of environment can be found in the Americas, from icy Arctic wastes to hot and humid rain forests. Since different habitats provide food and shelter for different kinds (species) of animals, this variety means that a huge range of species lives in North and South America, from top predators such as gray wolves and American alligators to tiny hummingbirds, frogs, and invertebrates.

While it is true to say that the landscapes of the Americas still support thousands of species, many are threatened by changes in land use, especially the clearance of large areas of forest and the conversion of natural grasslands to agriculture. Human disturbance of sensitive animals—for example, marine turtles on the Mexican beaches where they lay their eggs—is also a problem. And pollution caused by oil spills, such as those off the Galápagos Islands and in the Gulf of Mexico, kill animals and damage ocean environments. However, most governments now take measures to protect their wildlife and have established protected areas where development is restricted.

## Varied Habitats

In the far north, much of Alaska and northern Canada is covered with tiny shrubs and herbs; this landscape is called tundra. Despite cold temperatures and limited sun for much of the year, it supports many breeding birds. They arrive from warmer regions in spring to breed when there is an abundance of insect life to eat, before departing in fall. Mammals such as polar bears, caribou, and musk oxen also live there. Coastal waters provide feeding for walruses and whales. Further south, there are vast areas of coniferous forest called taiga, which are home to grizzly bears, elk, and deer, in addition to grouse, owls, and many other birds.

In the eastern United States, large tracts of deciduous forest remain. Red foxes, gray wolves, flying squirrels, and millions of song birds live in these forests. Some of these animals need large areas of unbroken woodland and these species have declined as forest has been broken up by highways and other developments. To the west, the Great Plains stretch for hundreds of miles to the foothills of the Rocky Mountains. These grasslands (shortgrass prairie in the west and tallgrass prairie farther east) were once home to huge numbers of American buffalo, or bison, but large-scale hunting in the late 19th century reduced the population of these large mammals from more than 60 million to just a few thousand. Protection in preserves has since allowed their numbers to increase but to nowhere near their former status. Much of the natural grassland of the Great Plains has been plowed for agriculture. However, the Great Plains still have more than 80 species of mammals, including black-tailed prairie dogs, 450 different kinds of birds, 57 species of snakes, lizards, and turtles, and some 15,000 insect species. In the southwest United States and northern Mexico, huge dry deserts such as the Mojave, Chihuahua, and Sonora are important for many reptiles, birds, and even amphibians that are able to take advantage of occasional rain.

The mountainous spine of Central America, running from Mexico to Panama, was once covered by mostly damp forest, which supported a great variety of animal life, especially amphibians and birds, but also monkeys, bears, and other mammals. Much of this forest has

been cleared for agriculture, sometimes resulting in erosion of the land as well as the destruction of habitats for the animals. Elsewhere, however, national parks protect the remaining forest. Monteverde Cloud Forest in Costa Rica is an important example, although protection here was not enough to save the golden toad from extinction.

The forests and high grasslands of the tropical Andes Mountains, running from western Venezuela through Colombia, Ecuador, Bolivia, and Peru, contain one-sixth of all plant species on Earth, despite covering only 1 percent of its land surface. Forest cloaks the lower slopes on both sides of this mountain range, becoming damp cloud forest and elfin forest (forest with very small trees) in the area just below the tree-line. Above the tree-line is open grassland, called *páramo*. Each of these habitats is home for different sets of mammals, birds, amphibians, and invertebrates, many

**The Florida Everglades** *provide a home for several endangered species, including American crocodile, Florida panther, and wood stork.*

## National Parks in the United States

Many areas of North and South America are now protected as national parks. This means that there are strict limits on any changes that can be made to the natural environment. It may be forbidden to cut down trees, drain wetlands, or build towns on areas of grassland, for example. Some preserves cover shorelines and areas of oceans, giving protection to whales, fish, and nesting turtles. In the United States, the National Park Service was set up in 1916 in an effort to preserve wildlife habitats for the enjoyment of future generations. However, the world's first national park, at Yellowstone, was established as long ago as 1872 to protect, among other things, populations of American bison, bobcats, mountain lions, beavers, sandhill cranes, and cutthroat trout. Yosemite National Park in California straddles the Sierra Nevada Mountains and has more than 250 species of fish, amphibians, reptiles, birds, and mammals. In the same state, Death Valley National Park is one of the hottest and driest places on Earth. This desert preserve has herds of bighorn sheep, as well as bobcats, mountain lions, and many snakes and lizards—and a surprising number of different species of birds. In contrast, the Everglades National Park in Florida is very wet, with an ample supply of fresh water and important populations of water birds, snapping turtles, American alligators, and Florida panthers.

of them rare. More than 1,700 bird species live there, almost one in five of all those found on Earth; 600 of these are found nowhere else. This is also the most important area on the planet for amphibians, with 980 species known, most of them found in the Andes and nowhere else. There are surely many more species of frogs and salamanders yet to be discovered. And there are probably thousands of invertebrates that scientists have yet to describe. In some areas, deforestation has been a problem, although forests on the steepest slopes have mainly been left untouched.

### The Amazon

To the east, the Amazon rain forest is probably the single most important region for animals anywhere on Earth. It is as large as the area between the Rocky Mountains and the Atlantic Ocean and its trees are home to a bewildering array of invertebrates, birds, and mammals. Fish and amphibians live in its mighty rivers, including the Amazon itself, while reptiles occupy river banks and the forest itself. However, large swaths of the forest have been cleared for highways, human settlements, and cattle ranches. This deforestation accelerated after 1972 when the Trans-Amazonian Highway was opened, allowing developers to reach deeper into the forest. Between 1991 and 2000 the total area of Amazon forest that had been cleared increased from 160,000 square miles (415,000 $km^2$) to 227,000 square miles (587,000 $km^2$), roughly the size of the Canadian province of Manitoba. Most of this forest was replaced with pasture for cattle, and many species of wild animals lost their homes. The total area of forest is now thought to be only four-fifths of its size in 1970. The good news is that the rate of clearance slowed dramatically after 2004, and there are important protected areas, such as Tumucumaque National Park in northern Brazil, which is the biggest area of protected tropical forest in the world.

Not all of South America is covered by forests. Also important are the continent's wetlands. Those of the Llanos in Venezuela are flooded in the wet season, when they are important for giant anteaters, caimans, anacondas, and many wading birds. Farther south, in Brazil, the Pantanal wetlands also have wading birds and several species of cats. Grasslands are also important. South of the Amazon rain forest, where the climate is too dry for forest growth, is the *cerrado* of Brazil, with animals such as giant armadillo, maned wolf, jaguar, puma, and many reptiles. In Argentina, the great expanses of the pampas grasslands support endangered pampas deer and pampas cats.

**San Rafael waterfall** *in Ecuador is surrounded by rain forest that is home to animals both common and endangered.*

EX
EW
CR
EN
VU
NT
LC
O

# Giant Anteater

## *Myrmecophaga tridactyla*

*Populations of the largest species of anteater have been in decline for much of their recorded history. Loss of habitat has been the most significant threat.*

The giant anteater leads a highly specialized way of life. It feeds mainly on colonies of large ants, which it collects with its specially adapted mouth and tongue. However, when it tracks down an ant nest teeming with potential food, the animal does not dive in and gorge. Instead, it breaks open a small section of the nest using its powerful 4-inch- (10-cm) long claws, dips its long snout into the hole, and extracts a couple of hundred ants with its long, bristly tongue. Often, after less than a minute's feeding time, it moves on, leaving the remaining ants to repair the damaged nest wall. In this way the ant colony continues to prosper, and there is always plenty of food when the anteater (or one of its neighbors) returns to the nest to feed again. The giant anteater has thus developed a sustainable way of exploiting its major food source.

A single anteater may have a home range of several thousand acres and will travel widely in the course of a day or night (anteaters can be active at any time, depending on the season and the weather). The ranges of neighboring anteaters overlap considerably, but the animals rarely meet; and if they do, they seem to ignore each other. Giant anteaters leave scent marks, but they use them more as a means of communication than as a territorial warning.

### Threatened Existence

Violence between anteaters is rare, and they usually choose to flee from danger rather than to fight it. Nevertheless, they are more than capable of defending themselves against attack from other animals. The anteater's main enemies, other than people, are jaguars and pumas. However, even for such accomplished predators an attack on an anteater is no easy undertaking. The anteater's long claws and hugely powerful forelegs are formidable weapons, capable of inflicting serious injury or even killing with a suffocating squeeze. The anteater's armory is no match for a shotgun or rifle, however, and unregulated hunting by trophy hunters has had a significant effect on the population.

The anteater's claws are so long that the animal has to walk on its knuckles and the outsides of its wrists to prevent its claws from scraping on the ground. Anteaters can move surprisingly well in this way, but they are not able to run fast. Their lack of speed makes them vulnerable to hunters and, at times, to forest fires: The anteater's long, shaggy fur is highly flammable, so there are always dead animals after a major fire. To compound the problem, breeding rates are slow, and there is a prolonged period of care for the young, so populations take a long time to recover from any losses.

**Giant anteaters** *are easily recognizable by their size and the distinctive dark stripe with white edges that runs from the chest onto both flanks. Their individual hairs are curious, being banded in black and white. A mother carries her single pup on her back for the first few months of its life.*

## DATA PANEL

**Giant anteater**

***Myrmecophaga tridactyla***

**Family:** Myrmecophagidae

**World population:** Unknown; possibly tens of thousands

**Distribution:** Central and South America, from southern Belize and Guatemala to Argentina, with populations in Brazil, Colombia, Costa Rica, French Guiana, Peru, Paraguay, and Venezuela

**Habitat:** Low-lying grasslands and open forests; sometimes ventures into dense vegetation

**Size:** Length head/body: 39–47 in (100–120 cm); tail: 26–36 in (65–90 cm); males 10–20% bigger than females. Weight: 39–86 lb (18–39 kg)

**Form:** Bulky body; distinctive tapered narrow snout. Gray-brown fur, long on legs and tail; white-bordered dark band tapers from chest to flank over each shoulder. Eyes and ears small. Long, narrow tongue can be extended up to 24 in (60 cm)

**Diet:** Ants, termites and their eggs, grubs, and cocoons; occasionally beetle larvae and fruit

**Breeding:** Single young born at any time of year after gestation of 6–7 months; fully weaned at 6 months, but may remain with mother for up to 2 years; mature at 3 years. Life span up to 26 years in captivity

**Related endangered species:** No close relatives, but other species of South American anteaters (for example, pygmy anteater, *Clyclopes didactylus*) are scarce and becoming more so with deforestation

**Status:** IUCN VU

The giant anteater depends for its survival on large areas of undisturbed habitat with plenty of suitable food. It does well in national parks and reserves, but elsewhere it is becoming scarce. Plowing for crops destroys ant colonies and inevitably affects the anteater's food supplies. The apparently wide distribution of the giant anteater—from Central America to northern Argentina—is misleading: Within this range populations are small, spread out, and often isolated in small "islands" of suitable habitat. The species is not yet at the brink of extinction; but unless steps are taken to secure its habitat, the future of the giant anteater will be in crisis.

EX
EW
CR
EN
VU
NT
LC
O

# Giant Armadillo

## *Priodontes maximus*

*South America's giant armadillos are solitary creatures with a large home range. Human activities, such as logging and the spread of urban settlements and farmland, are now driving the prehistoric-looking mammals toward extinction.*

The giant armadillo is one of 21 armadillo species, which include the nine-banded armadillo, the only one whose range extends into the United States. Armadillos have a protective armor of bony plates that grow in bands over the head, back, and hips. Their long, curved claws are good digging tools, and they are strong swimmers. They can also swallow air to make themselves buoyant in water.

The giant armadillo is the largest of the species. Humans aside, it has few natural enemies, and its size and body armor equip it well against potential predators such as the jaguar. Its 80 teeth (more than any other land mammal) do not need to be used for defense, and they are small and set far back inside the armadillo's mouth.

The giant armadillo's range covers a huge area from Venezuela to Argentina, but between these points its distribution is patchy, and the animals are thinly spread. This is partly because of the armadillo's solitary lifestyle. Adults tend to avoid each other, except during the breeding season. Nocturnal creatures, they rest in burrows during the day and spend the night patrolling their territory. They feed on termites, ants, and other insects as well as snails, small snakes, and carrion. They leave smelly scent marks to make sure their neighbors know that they are around.

Their extensive home ranges mean that a healthy population of giant armadillos needs a large area of continuous habitat. Small patches of forest are of no use, since the giant armadillos are sensitive to disturbance and will not venture far beyond the cover of trees in order to feed.

### Prime Habitat

Large areas of prime giant armadillo habitat are being

## DATA PANEL

**Giant armadillo**

***Priodontes maximus***

**Family:** Dasypodidae

**World population:** Unknown; perhaps tens of thousands

**Distribution:** South America, from Venezuela to northeastern Argentina, including Colombia, Guyana, Surinam, French Guiana, Peru, Brazil, Bolivia, and Paraguay

**Habitat:** Forests and scrublands, especially those close to water; occasionally ventures into grassland, but never goes far from tree cover

**Size:** Length head/body: 30–39 in (75–100 cm); tail: 18–20 in (45–50 cm). Weight: 99–132 lb (45–60 kg)

**Form:** Bulky, dark-brown animal with long tail; head, back, and hips covered by rows of small, bony, horn-covered plates (called scutes); elsewhere skin covered with stiff hair. Snout slightly elongated with long, flexible tongue; eyes and ears small. Front feet have long, curved claws

**Diet:** Mostly termites; also ants and other insects; occasionally worms, spiders, snails, and small snakes; some carrion

**Breeding:** One, or occasionally 2, young born after 4-month gestation; weaned at 4–6 weeks; mature at 9–12 months. May live up to 15 years in the wild

**Related endangered species:** Pink fairy armadillo *(Chlamyphorus truncatus)* DD; greater fairy armadillo *(Chaetophractus retusus)* DD; Brazilian three-banded armadillo *(Tolypeutes tricinctus)* VU; others poorly known

**Status:** IUCN VU

destroyed continuously by logging and the spread of human settlements and agriculture. The deliberate flooding of large areas of rain forest to create reservoirs is also threatening armadillos, along with many other kinds of South American wildlife.

Giant armadillos rapidly disappear from the areas around new human settlements or roads. It may be that some simply choose to move on, retreating back into less disturbed forest; but it is more likely that many are shot for their meat, which is a local delicacy.

## Hidden Possibilities

Armadillos have the unlikely characteristic of being the only known hosts—apart from human beings—of leprosy. The disease is usually transmitted to people through the soil and causes lumps, discoloration of the skin, and gradual deterioration of the nerves, especially in the extremities such as the fingers and toes. Leprosy is still a major disease in many parts of the world, so armadillos may be an important model for scientists in their efforts to find an effective treatment.

## Giant Ancestors

The giant armadillo is not very large compared with its ancient forebears. Fossils show that South America was once home to enormous armadillos called glyptodonts whose body plates were fused to form a domed carapace (thick, hard shield) like that of a tortoise. They were over 8 feet (2.5 m) long—about the size of a small car. These giants seem to have disappeared during the last 2 million years. It is now up to us to ensure that the modern-day giant armadillo does not go the same way.

**The giant armadillo's** *armor-plated hide provides it with excellent protection from potential predators, which allows it to go about its termite-eating business in peace. Unlike some other armadillos, this one cannot roll up completely, but it often stands up on its hind legs.*

EX
EW
CR
EN
VU
NT
LC
O

# Arizona Black-tailed Prairie Dog

## *Cynomys ludovicianus*

*Although not yet an endangered species, the Arizona black-tailed prairie dog has suffered huge losses of habitat and now survives mainly in protected areas. Other prairie dog species face similar but potentially more serious problems, since they never breed as quickly as their black-tailed cousins.*

The prairie dog is not a dog at all, but a type of ground squirrel. It is a colonial creature that lives in large numbers in extensive burrow systems called towns. In fact, "cities" might be a better description. Early explorers described some prairie-dog towns as extending for many miles and containing tens of thousands of animals. The largest town ever recorded was said to cover 25,000 square miles (65,000 sq. km) and may have been home to 400 million prairie dogs.

Within the towns the animals live in small social groups called coteries, each occupying about 1 acre (0.4 ha). Coteries consist of several males and females, generally close relatives, together with their offspring. They defend their part of the town against rival groups. Prairie dogs have a very complex social life, involving communication with others and complicated behaviors to prevent inbreeding.

Burrow entrances are sited between carefully shaped mounds, the position of which allows a constant flow of fresh air to enter the underground tunnels. Other burrows have mounds at the entrance, providing the prairie dogs with vantage points from which they can survey their territory. From the mounds lookouts keep constant watch for danger and for trespassers. The animals often emit a loud double bark, throwing their front legs into the air and arching their back as they stand upright on their hind legs. This keeps them in contact with the rest of the community. Vocal warnings alert others to the approach of predators such as coyotes and badgers. Aerial predators such as hawks and eagles are also a threat and sometimes cause the sentries to call in a slightly different way.

### Making Enemies

The prairie dogs nibble down taller vegetation to maintain a short turf throughout their town, making

### DATA PANEL

**Arizona black-tailed prairie dog**

*Cynomys ludovicianus*

**Family:** Sciuridae

**World population:** Up to 18 million

**Distribution:** Western central U.S., from Canadian to Mexican borders

**Habitat:** Open shortgrass prairies

**Size:** Length head/body: 11–13 in (26–31 cm), female about 10% smaller than male; tail: 3–4 in (7–9.5 cm). Weight: 20–52 oz (575–1,490 g)

**Form:** Short-legged ground squirrel, with sandy-brown coat and black tip to tail

**Diet:** Mainly grasses but also some other plants; occasionally grasshoppers

**Breeding:** Three to 5 young born March–April after gestation period of about 4 months. Life span about 5 years

**Related endangered species:** Mexican prairie dog *(Cynomys mexicanus)* EN; Utah prairie dog *(C. parvidens)* EN

**Status:** IUCN LC

it difficult for predators to approach without being seen. It also means that they destroy some of the local vegetation. Thousands of prairie dogs can eat a great deal of grass and other prairie plants, which makes them unpopular with ranchers, who see good grazing material nourishing prairie dogs instead of beef cattle. One prairie dog may not eat very much, but even the animals in a small prairie-dog town will consume as much as several cattle in a day. The many burrow entrances and shallow tunnels create a dangerous terrain for horses and their riders. Horses risk broken legs as they canter or gallop across a town, and their riders may be thrown off and seriously injured. Ranchers have gone out of their way to get rid of prairie dogs, usually using poison that is heaped into the burrows and kills large numbers of dogs cheaply.

Elsewhere, the prairies have been turned into farmland. Farmers want to get rid of the prairie dogs because their burrows ensnare tractors and obstruct the planting and harvesting of crops. Deep plowing destroys the prairie-dog towns anyway, and the new crops create an unsuitable environment for the animals.

## Wiped Out

As a result of these pressures and government policies designed to protect the rangeland, prairie dogs have been wiped out over large areas. The Arizona black-tailed prairie dog is found mainly in protected areas, such as Wind Cave National Park in South Dakota. The closely related Utah prairie dog, which was never common anyway, has disappeared from about 90 percent of its former range, and the Mexican prairie dog is now endangered.

**Prairie dogs** *often stand on their hind legs to get a better view. They are always on the lookout for predators.*

# Golden Lion Tamarin

## *Leontopithecus rosalia*

*Most of the coastal forests in which the golden lion tamarin lives have been felled. Reintroductions of captive-bred animals have focused attention on the need to conserve the remaining habitat.*

The tropical forests of Brazil's Atlantic coast used to be rich in wildlife, but since the early 19th century over 90 percent of the forest has been felled for timber and for fuel or to create space for people and agriculture. The forests have always been the only home of the golden lion tamarin, whose habitat was eventually reduced to barely 350 square miles (900 sq. km) and divided into more than a dozen separate patches. To make matters worse, many of these attractive creatures were captured for zoos or for sale as pets.

Golden lion tamarins are social animals that breed in small groups of about five individuals. All members of a group help care for the few young that are produced by the breeding females. If the group is disturbed or reduced to only two or three animals, the survivors will inevitably be less successful in raising young. Any reduction in numbers therefore leads to a downward spiral of ever fewer animals with even lower breeding success.

Numbers of the species as a whole fell to a critically low level in the 1970s, when the golden lion tamarin became one of the world's rarest mammals. A captive-breeding program involving zoos in several countries was begun in 1973. Within 10 years the number of animals involved had risen from about 70 to nearly 600, enabling some to be taken back to Brazil for reintroduction into the wild. The release project began in 1984, when 15 animals were set free in the coastal forest. Several soon died from disease, snakebite, or dog attack, but the first baby was born within a few weeks.

### DATA PANEL

**Golden lion tamarin**

***Leontopithecus rosalia***

**Family:** Callitrichidae

**World population:** More than 1,000

**Distribution:** Coastal forest of Brazil

**Habitat:** Lowland tropical forest from sea level to about 3,000 ft (914 m)

**Size:** Length head/body: 8–13.5 in (20–34 cm); tail: 12.5–15.8 in (32–40 cm). Weight: 21–28 oz (600–800 g)

**Form:** Tiny monkey with long, silky golden hair and a long tail

**Diet:** Mostly fruit and insects; also small animals; occasionally birds' eggs

**Breeding:** Average of 2 young born September–March each year after 4-month gestation; mature at 2–3 years. Life span up to 15 years in captivity, less in the wild

**Related endangered species:** Black-faced lion tamarin *(Leontopithecus caissara)* CR; golden-headed lion tamarin *(L. chrysomelas)* EN; golden-rumped lion tamarin *(L. chrysopygus)* EN

**Status:** IUCN EN

### Back to the Wild

Initially the released tamarins found it difficult to cope with the challenges of survival in the wild. Used to being cared for by humans, they had lost crucial food-gathering skills. In recognition of the problems they faced, scientists took care to give subsequent animals some experience of conditions in the wild before releasing them.

A period of adjustment helped improve the reintroduction program's success rate. By the early 1990s golden lion tamarins that had been released into the wild were breeding more successfully than their cousins in captivity. It was anticipated that in time the wild population would no longer be dependent on humans to provide extra food and protection.

The population has now risen above 1,000 and is still centered on the Poco das Antas Reserve in Rio de Janeiro Province. As numbers have built up, and more animals have become available from zoos, additional populations have also been established in other parts of the coastal forest. Spreading the population reduces the risk of disease or fire wiping out the whole species at once.

Tamarin social groups are territorial, each needing an exclusive range of about 100 acres (40 ha). Tiny remnant patches of forest are consequently of little use to the animals and do not allow the population to expand. It therefore becomes vital to link up habitat patches by planting more trees. Thanks to conservation efforts since the project began, there has in fact been a 38 percent increase in forest cover in the areas where the tamarins live. Replanting is also bringing benefits to many other species.

Fortunately, local people—including major ranch owners in the region—have adopted the cause of the golden lion tamarin and are proud of the part they have played in helping conserve the species. An educational program has involved schoolchildren in the campaign to protect the animal's future.

The success of the golden lion tamarin conservation project has had the added benefit of focusing attention on the plight of other tamarin species. In 1998 Brazil's Superagui National Park was extended by more than 50 percent to accommodate the endangered black-faced lion tamarin. Several zoos are now breeding tamarin species that may one day be released to supplement wild populations elsewhere in South America.

**Golden lion tamarins** *are so called on account of their long mane of hair. All four species have a golden coloration, but only the golden lion tamarin is golden from head to tail.*

# Cuban Solenodon

## *Solenodon cubanus*

*The Cuban solenodon is a primitive animal. Always scarce, its numbers were drastically reduced by the activities of European settlers and exposure to new predators such as dogs.*

Solenodons are ancient creatures. They resemble some of Earth's earliest mammals, those that evolved millions of years ago. Their closest relatives are now mostly extinct, but the modern solenodon has managed to carve out an existence on the island of Cuba. A similar species is found on the neighboring island of Hispaniola (now divided between Haiti and the Dominican Republic). The Hispaniolan species is also endangered.

The animals are mainly nocturnal. They feed by rooting around in leaf litter and turning over stones and rotting wood with their long, probing snout. They extract grubs and worms by tearing open rotting wood with the sharp claws on their forefeet.

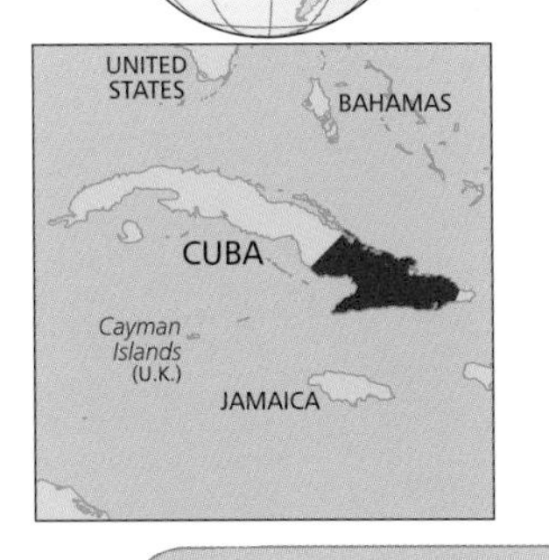


Solenodons live alone, passing the day in a den among logs or rocks. They particularly favor areas of coraline limestone where there are many nooks and crannies in which to hide. An added attraction is the invertebrate fauna of snails and beetles that lives among the rocks to avoid the hot, dry conditions in the open—and provides food.

Like its relatives, the Cuban solenodon is a relatively slow-moving animal that seems to walk with a drunken stagger. Although it can climb well, it is not able to jump, so it is easy game for predators.

### Habitat Invasion

Solenodons breed slowly, having only a few young at a time, with long intervals in between. As a result, solenodon populations are not able to withstand heavy losses. The main problem for the Cuban solenodon has been the colonization of its island home by European settlers. The settlers introduced cats and dogs, and later mongooses, to control rats in the sugar plantations. Such predators were previously absent from the island, so the species had evolved safely

### DATA PANEL

**Cuban solenodon**

***Solenodon cubanus***

**Family:** Solenodontidae

**World population:** Unknown; probably low hundreds

**Distribution:** Eastern Cuba

**Habitat:** Dense, damp jungles; also thick scrub and around plantations

**Size:** Length head/body: 11–13 in (28–32 cm); tail: 7–10 in (17–25 cm). Weight: about 1.5–2.2 lb (0.7–1 kg)

**Form:** Resembles giant shrew with long, pointed snout and long, bare tail. Usually dark brown or blackish. Five toes on each foot; prominent claws on toes. Grooved lower incisor tooth carries toxic saliva to prey when bitten

**Diet:** Mainly insects and small animals (lizards and spiders, for example); also plant material

**Breeding:** One or 2 young per litter; young stay with mother for several months (a long time for an insectivore). May live up to 6 years

**Related endangered species:** Hispaniolan solenodon *(Solenodon paradoxus)* EN

**Status:** IUCN EN

without them. Needless to say, the solenodons' small teeth were no match for a cat or large dog, and individuals were caught in large numbers by these new predators.

Moreover, all the islands in the Caribbean have been under pressure from development, involving clearance of huge areas of subtropical forest. Much forested land has been converted into farms and sugar plantations. Farming activity has deprived the solenodons of their main habitat and feeding places. In addition, the open terrain created by farmland makes solenodons more vulnerable to predation, since they have nowhere to hide. As a result, solenodons have become extremely rare. In fact, by the 1960s the Cuban species was believed to be extinct, since none had been reported since 1890.

## Legal Protection

Surveys in the 1970s suggested that the plight of the solenodon was not as bad as had been supposed, however. Scientists located some living specimens, and it was concluded that although the animals were rare, they were probably widely distributed in several parts of Oriente Province in eastern Cuba. Although Oriente Province takes up only a small part of the island, the human population there is comparatively low, so the solenodon is less at risk from habitat disturbance and predation by domestic animals. Another advantage for the Cuban solenodon is that it is not considered to be of economic or medicinal importance. As a result, there has been little pressure on its numbers from activities such as hunting or trapping. The Cuban solenodon has been given full legal protection. Although the last live animal was seen as long ago as 2003, it is hoped that the species will benefit from protected areas in the eastern Cuban highlands that have been specially set aside for the conservation of wildlife.

**Solenodons** *resemble giant shrews. However, despite their size, they are poorly equipped to defend themselves against introduced predators such as cats and dogs.*

EX
EW
CR
EN
VU
NT
LC
O

# Hawaiian Monk Seal

## *Monachus schauinslandi*

*Until the Hawaiian Islands were colonized by humans, monk seals bred on most of the islands and atolls, taking advantage of the predator-free beaches to rear their pups. Today the story is different.*

The Hawaiian monk seal's range started to contract when the main islands of Hawaii were colonized by Polynesians in 400 CE. For a time monk seals still thrived on the smaller atolls and islands in the north of the archipelago, where they were rarely disturbed. However, it was not long before even these remote islands were also discovered, and the seal's situation deteriorated rapidly.

Despite living in warm, tropical seas, the Hawaiian monk seal has fur as luxurious as that of other seals. In the early 19th century hunters killed so many Hawaiian seals that by 1824 the species was thought to be extinct. In fact, small breeding populations had survived on some of the more remote Hawaiian islands. When such groups were discovered, however, they too soon became prey to hunters.

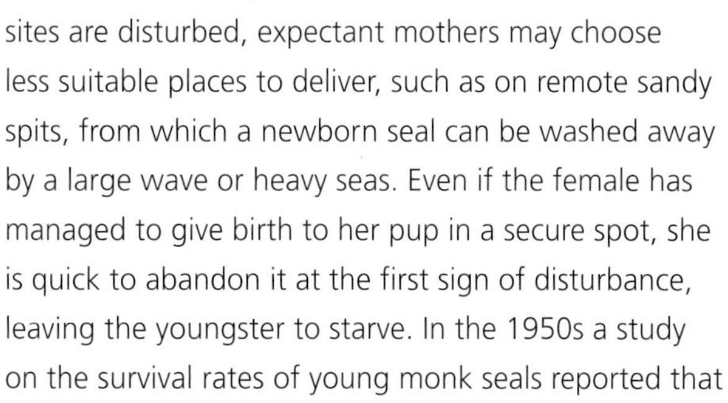

Along with hunting, the mere presence of humans has also affected monk seal populations. The animals are extremely sensitive to disturbance, especially during the spring, when females give birth on sandy beaches. Under ideal circumstances the pups are born at the top of the beach, well above the high-tide mark and in the shelter of beachside vegetation. However, if the seal's traditional breeding sites are disturbed, expectant mothers may choose less suitable places to deliver, such as on remote sandy spits, from which a newborn seal can be washed away by a large wave or heavy seas. Even if the female has managed to give birth to her pup in a secure spot, she is quick to abandon it at the first sign of disturbance, leaving the youngster to starve. In the 1950s a study on the survival rates of young monk seals reported that

### DATA PANEL

**Hawaiian monk seal**

***Monachus schauinslandi***

**Family:** Phocidae

**World population:** About 1,100

**Distribution:** Mostly restricted to the Leeward Islands north of Hawaii

**Habitat:** Shallow lagoons and sandy beaches

**Size:** Length: 6.6–8.2 ft (2–2.5 m); females around 10% longer than males. Weight: 330–660 lb (150–300 kg); females heavier than males

**Form:** Dark-gray seal with velvety fur, silvery on underside; face has long, dark whiskers

**Diet:** Reef fish and invertebrates, especially eels and octopuses

**Breeding:** Single pup born every 1 or 2 years, usually between March and May; weaned at 6 weeks; mature at 5–6 years. May live as long as 30 years

**Related endangered species:** Mediterranean monk seal *(Monachus monachus)* CR; Caribbean monk seal *(M. tropicalis)* EX (last seen in 1952)

**Status:** IUCN CR

of all the pups born on a particular stretch of coast, 39 percent died before they reached the sea.

As more Hawaiian islands were exploited in the 20th century, the monk seals retreated. In 1978 up to 60 seals on Laysan Island died suddenly from a form of food poisoning known as ciguatera. The seals had eaten prey contaminated with a single-celled organism that produces a potent toxin. Such an event might seem like a natural hazard of a seafood diet, but it is now accepted that outbreaks of ciguatera are nearly always associated with damage to coral reefs. The incident was probably triggered by destruction of the reef during work on the harbor around Midway Atoll.

Generations of Hawaiians have made their living from the sea, and today commercial fisheries pose yet another threat to the seals. Every year seals are killed accidentally or deliberately: Debris, oil, and diesel spilled from all kinds of ships present a problem for the seals and other endangered wildlife, and many drown after becoming entangled in fishing nets.

**Monk seals** *have lived on the remote islands of Hawaii for about 15 million years. In that time seals in other parts of the world have evolved and adapted, while the isolated Hawaiian population has hardly changed at all.*

### A More Secure Future

In 1980 the United States Fish and Wildlife Service took action. A monk seal recovery team was formed to safeguard the remaining population. Today the entire breeding range of the Hawaiian monk seal falls within a specially designated national wildlife refuge, and the seals are strictly protected. A feature of the population is that there are three times as many males as females. Competition among the males is intense: Sometimes newborn pups are crushed or females killed by excited bull seals. In a small population such losses are significant. By removing some of the males, and giving others a testosterone-inhibiting drug to dampen aggressive behavior, the recovery team hopes to give the population a more secure future.

# Northern Fur Seal

## *Callorhinus ursinus*

*Once abundant, the northern fur seal was commercially hunted under international agreement. Recently, numbers have fallen dramatically for reasons that are not understood.*

Northern fur seals live in the north Pacific between Kamchatka (Siberia) and Alaska. There was once thought to be a total of about 4 million, but uncontrolled hunting reduced their numbers severely. Their thick, densely furred skin was highly prized for warm garments, and specially prepared furs were at one time very fashionable.

In the past hunters would shoot large numbers of northern fur seals out at sea. Not only was the method cruel, it was also extremely wasteful: Many seals were wounded, and the bodies were never recovered. Hunters soon discovered that it was easier—and more efficient—to kill the animals on the beaches where, for eight weeks of the year, they come ashore to breed.

Northern fur seals breed on the Pribilof Islands (off Alaska) and the Commander Islands (east of Kamchatka), and previously were also found in small numbers on Japanese territory. Each year they migrate to the gloomy, rain-soaked shores, traveling up to 6,200 miles (10,000 km) to breed. Between 1867 and 1911 more than 1 million were taken, with at least as many again being killed and lost at sea.

It has proved extremely difficult to protect wildlife living in international waters. Attempts to control exploitation of seals in the early 20th century failed, and the population dwindled; some colonies died out altogether. In 1911 it was agreed to make the killing of fur seals illegal everywhere except the Pribilof colonies, which were tightly controlled by the United States. The American government managed the herds and shared the proceeds with its partner nations. The strategy seemed to work well. From a low point of fewer than 250,000 animals, the population grew to over 3 million by 1940. The increase occurred despite an annual harvest of about 40,000 seals.

### DATA PANEL

**Northern fur seal**

***Callorhinus ursinus***

**Family:** Otariidae

**World population:** About 1.1 million

**Distribution:** North Pacific coasts as far south as California; main breeding on Pribilof Islands off Alaska and Commander Islands off Kamchatka

**Habitat:** Open sea within 60 miles (100 km) of the coast; comes ashore to rocky beaches to breed

**Size:** Length: male up to 6.5 ft (2.1 m); female up to 4.6 ft (1.2–1.5 m). Weight: male 300–615 lb (136–279 kg); female 66–110 lb (30–50 kg)

**Form:** A large fur seal; bulls reddish brown and black, cows pale and gray

**Diet:** Mainly fish

**Breeding:** One young born per year; mature from about 4 years. Life span probably more than 30 years

**Related endangered species:** Steller's sea lion *(Eumetopias jubatus)* NT; Galápagos fur seal *(Arctocephalus galapagoensis)* EN; Juan Fernandez fur seal *(A. philippii)* NT; Guadaloupe fur seal *(A. townsendi)* NT

**Status:** IUCN VU

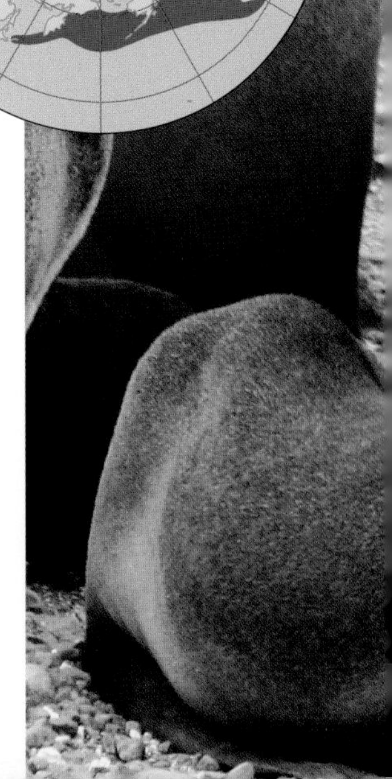

### Conservation Success

By the mid-20th century the case of the fur seal had become an example of successful conservation management combined with sustainable harvesting of animals in the wild. The harvesting, or culling, was possible because the northern fur seal is a harem breeder. The adult bulls come ashore in early summer and defend a breeding territory, keeping younger males at bay. Each "master" bull then has up to 100 females (cows) to himself, each of which is only a quarter of his size. This is the greatest male–female size difference of any mammal.

Only a small proportion of male fur seals ever breeds. The rest are kept away from the breeding beaches by the territorial bulls and can be found gathered together nearby in all-male groups. Many of the surplus males can therefore be killed with virtually no effect on the breeding population. The culling may even benefit the rest of the population by making more food available to those that are left, including females and young.

### Unaccountable Decline

The harvest was managed in this way until 1984, the year when commercial hunting stopped. Although seals are now protected and no longer culled, their numbers have been in decline for many years: The main breeding population has halved in less than 50 years. Records show that about 50,000 seals drown in fishing nets each year, although this is not enough to account for the long-term decline. One explanation may be that overfishing has reduced the seal's food supply to a level that cannot support the previous numbers of seals.

There is better news from the Russian colonies, where fur seal numbers have held up: Some groups have recently recolonized Robben Island in the Sea of Okhotsk. And about 8,000 animals now breed on San Miguel Island off southern California.

**Northern fur seals** *come ashore to rocky beaches to breed. Harems of between 40 and 100 females are dominated by one adult bull. Females, or cows (below), do not have the heavy mane of the bulls.*

EX
EW
CR
EN
VU
NT
LC
O

# Grizzly Bear

## *Ursus arctos*

*The grizzly bear has a fearsome reputation and is likely to come off best in any fight. Through competition with people, however, it has lost habitat, and it is still threatened by hunting.*

Many apparently different types of grizzly bear have been described, but they are probably all the same species. Some have been recognized as subspecies, such as the Eurasian brown bear. The true grizzly bear has distinctive white-tipped or "grizzled" hairs around the face and a substantial hump that make it look quite different from other forms of the brown bear, although it is zoologically the same species.

Grizzly bears are successful animals. They are able to live in a variety of habitats, and they can eat almost anything. They have few natural enemies and were once spread across most of the Northern Hemisphere. Today they are still fairly numerous in Alaska and northern parts of Russia. Everywhere else they are either severely threatened or already extinct. In Britain, for example, the last grizzly bears died out about 1,000 years ago, while more recently grizzly bears have disappeared from China, Korea, and southern Japan. The Mexican race is Extinct, and the Great Plains race is Endangered.

### DATA PANEL

**Grizzly bear (brown bear)**

***Ursus arctos (U. horribilis)***

**Family:** Ursidae

**World population:** More than 200,000, including more than 100,000 in Russia, 33,000 in the United States, and 25,000 in Canada

**Distribution:** Across northern parts of Northern Hemisphere from eastern Europe to Japan; also Alaska, northwestern Canada, and the U.S.

**Habitat:** Forests, tundra, and mountains

**Size:** Length head/body: 6–9 ft (1.8–2.7 m); height at shoulder: 36–60 in (90–150 cm). Weight: male 330–695 lb (150–315 kg); female 132–440 lb (60–200 kg)

**Form:** Large bear, various shades of brown with humped shoulders

**Diet:** Fish, meat, fruit, grass, roots—almost anything edible

**Breeding:** Between 1 and 3 cubs born early in year, stay in den for 4 months; live with mother for at least 2 years. Life span over 30 years in wild; 50 years in captivity

**Related endangered species:** Mexican grizzly bear *(Ursus arctos nelsoni)* EX; Asian black bear *(U. thibetanus)* VU; sloth bear *(Melursus ursinus)* VU; spectacled bear *(Tremarctos ornatus)* VU; polar bear *(U. maritimus)* VU

**Status:** IUCN LC

### Perceived Threats

Like gray wolves, bears have a grim reputation in ancient folklore and were once widely feared. Even today, people are rightly wary of bears, although the animals are not persecuted to the degree they once were. All types of brown bear have suffered a similar fate; various races died out, at least in part, through persecution by humans.

Bears are big and can certainly be dangerous, especially when provoked or threatened. Mothers are extremely protective of their young cubs. In some North American national parks there have been serious and sometimes fatal attacks by bears on people. The bears are also known to raid crops and beehives. The knowledge of what bears can do, added to the fear of what they might do, has meant that for centuries they have been persecuted in order to remove a perceived threat. In the few parts of Europe where they survive, farmers protest at the presence of bears in their area. The practice of hunting and trapping bears continues, despite legal protection.

The size of the bears is not only a problem for people, but for bears too. Large animals tend to be naturally fewer in number, and their populations

**A grizzly bear** *in search of a meal. Young females may stay in the mother's range after leaving her care; males disperse widely.*

are more spread out across their habitats than smaller species. Even in ideal habitats, such as the Yukon Territory, Canada, or Yellowstone National Park in the United States, there may be only one bear in every 10 square miles (one per 26 sq. km); in parts of Europe there are even fewer—in Sweden there is one bear per 400 square miles (one per 1,000 sq. km).

Low-density populations are essential to reduce competition for food, but they also mean that the animals have to wander widely in search of mates. Most bears have huge home ranges, often 100 to 1,600 square miles (250 to 4,000 sq. km). They need a lot of space, but increasingly so do people. As human populations expand and need more land, the bears are forced into smaller ranges; their populations become fragmented. The bears end up in smaller groups and so are in danger of inbreeding or dying out.

National parks tend to support high-density populations—for example, one bear in every 3 square miles (one per 7 sq. km) in Abruzzo National Park, Italy. However, the bears still wander widely and often run into trouble when they leave the park. Bear numbers now seem stable and may be increasing. In an effort to boost the tiny population in the French Pyrenees, bears from Slovenia were released there in 1996 and 1997. Although they bred, by 2009 there were only 20. More releases are needed, but they have met stiff opposition from local farmers, afraid they will eat their sheep.

EX
EW
CR
EN
VU
NT
LC
O

# Red Wolf

## *Canis rufus*

*It has taken an immense conservation effort to save the red wolf from extinction; and while a small number have been successfully returned to the wild, the fight to preserve the species is by no means over.*

When Europeans first colonized North America, the red wolf was abundant from Texas to Pennsylvania in the north and to Florida in the south. For the settlers the wolves embodied all that was wild and frightening about the New World. Worried about dangers to their livestock and to themselves, people shot, snared, trapped, and poisoned the wolves without mercy.

At the same time, the wolves' wilderness habitat was tamed. Forests were cleared and prairies turned into pasture or plowed. The changes in the landscape brought a further threat to the wolves from their close relative the coyote. Coyotes are less well suited to forested landscapes and prior to interference by settlers had remained well to the west of the red wolf's habitat. However, as the forests disappeared, the coyotes moved in. The wolf and coyote were able to interbreed, so even in areas where persecution was minimal, purebred red wolves began to disappear, although hybrids remained.

By the 1960s the purebred red wolves were restricted to a small area of swampy coastal prairie on the Gulf coast of Texas and Louisiana. It was a poor-quality habitat, and the wolves struggled to survive. Over half the pups died from hookworm infections, and virtually every adult was afflicted with heartworm and mange. In 1975 the situation reached a crisis point. Conservationists working for the recently established Red Wolf Recovery Team believed that the only way to save the species was to bring the entire surviving population into captivity.

For a decade there was not a single red wolf left in the wild. Fourteen of the healthiest wolves were selected to breed in captivity. By the time the last of the wild-caught wolves died in 1989, small populations of captive-bred descendants had been reintroduced

### DATA PANEL

**Red wolf**

***Canis rufus***

**Family:** Canidae

**World population:** About 300 (2013 estimate)

**Distribution:** Most in captivity; small reintroduced populations on islands off Florida coast, Mississippi, and South Carolina; also on mainland in North Carolina

**Habitat:** Forest, swampland, and prairie

**Size:** Length head/body: up to 4.3 ft (1.3 m); tail: up to 17 in (42 cm); male larger than female. Weight: 40–88 lb (18–40 kg)

**Form:** Narrow-bodied, long-legged wolf with large ears. Coat varies from tawny to gray or black, tinged with red

**Diet:** Rodents, rabbits, deer, hogs, crayfish, insects, and carrion

**Breeding:** Up to 12 young (usually 3–7) born in late spring after 9-week gestation; weaned at 8–10 weeks; mature at 22–46 months. Life span up to 14 years

**Related endangered species:** Gray wolf *(Canis lupus)* LC

**Status:** IUCN CR

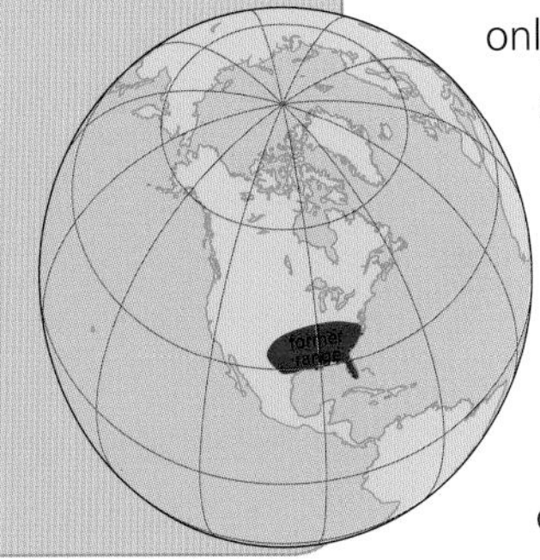

to several small island refuges in Florida and South Carolina, and to one mainland site, Alligator River Wildlife Refuge in North Carolina. In 2013 there were more than 100 red wolves living in the wild and nearly 200 in captivity.

## An Uncertain Future

Wolves divide public opinion, and the red wolf has its enemies as well as its fans. The antiwolf lobby claims that the reintroductions put livestock and people at risk. In 1996 much was made of scientific research that indicated that the red wolf was in fact a hybrid between the gray wolf and the coyote. The implication was that only true species should be entitled to protection. It is possible that prior to the changes that put red wolves and coyotes in the same habitat, the red wolf was on its way to becoming a true species. Conservationists believe that losing a species in the making is as bad as losing one that is established.

**The red wolf** *appears to be thriving where it has been reintroduced. Its ability to hunt and breed successfully does not seem to have been affected by a captive upbringing.*

EX
EW
CR
EN
VU
NT
LC
O

# Vicuña

## *Vicugna vicugna*

*Prized for centuries for its extremely fine wool, the vicuña has suffered from uncontrolled hunting in recent times. However, its future now seems more secure.*

The vicuña lives in small family herds of between five and 10 individuals led by one dominant male. Surplus males may form much larger, loosely organized herds of up to 150 animals. The vicuña is the smallest member of the camel family. Its home is the dry, grassy plains of the high Andean plateau, over 13,000 feet (4,500 m) above sea level.

Vicuñas feed during the day and retreat to a different area to spend the night, using a regular pathway. Unlike other members of the camel family, they need to drink regularly, and they are therefore rarely found far from water. Their territory is marked by communal dung piles, and the animals may range widely over more than 500 acres (200 ha).

Vicuñas are slender animals with a pale-brown coat, reddish-brown coloring on the neck and head, and off-white fur beneath. They are extremely wary and constantly on the lookout for danger. When threatened, they make a loud, high-pitched whistle as a warning call.

Vicuñas can run at speeds of up to 30 miles per hour (50 km/h), despite the high altitude, which would leave most animals breathless. The vicuña is assisted by an exceptionally efficient blood circulatory system. It has a powerful heart, which is larger than in other animals of a similar size, and its blood is capable of absorbing large quantities of oxygen from the air.

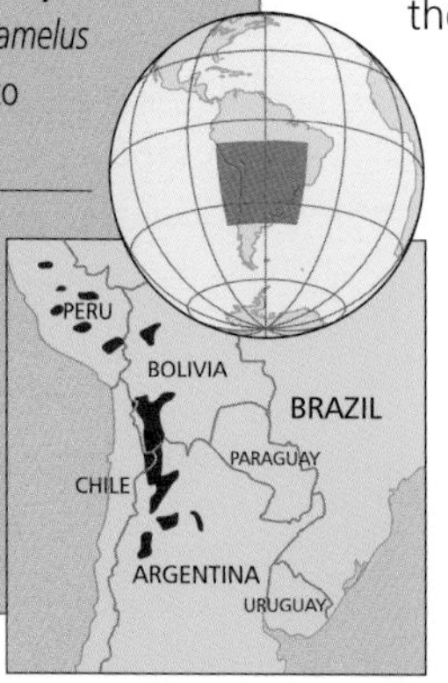

### A Precious Commodity

Like other mammals that live at altitudes of up to 19,000 feet (5,800 m), the vicuña is exposed to extremely low temperatures, especially at night. As protection it has developed a very long coat of fine wool, which traps a layer of warm air against the skin and reduces the amount of body heat lost.

The vicuña's fine fur has also been highly prized by people, who have long used it for making warm clothing. For centuries the native people of the high Andes would round up wild herds of vicuñas and strip them of their wool to spin and weave into a soft cloth that was worn by the nobility. Since the vicuña's molted their wool seasonally, it was also possible to drive the herds through dense vegetation and collect the wool scraped off by branches. Such activities rarely resulted in anything more than scaring the animals, so they probably had little effect on the species' abundance.

## DATA PANEL

**Vicuña**

***Vicugna vicugna***

**Family:** Camelidae

**World population:** About 127,000 (2008 estimate)

**Distribution:** Andes of southern Peru, western Bolivia, northwestern Argentina, and northern Chile

**Habitat:** Dry, grassy plains at least 13,000 ft (4,500 m) above sea level. Rarely found far from water

**Size:** Length head/body: 4–6 ft (1.25 m–1.9 m); tail: 6–10 in (15–25 cm); height at shoulder: 2.5–3.5 ft (70–100 cm). Weight: 80–140 lb (35–65 kg)

**Form**: Slender body and legs. Pale-brown coat; dirty white underneath. Hair forms a long, white mane on the chest

**Diet:** Almost entirely grass; needs water every day

**Breeding:** One young per year after 7-month gestation. Life span 15–20 years in the wild

**Related endangered species:** Wild Bactrian camel *(Camelus bactrianus)* CR; guanaco *(Lama guanaco)* LC

**Status:** IUCN LC

PERU
BOLIVIA
BRAZIL
PARAGUAY
CHILE
ARGENTINA
URUGUAY

The total number of vicuñas was once probably about 1.5 million. When the Spaniards invaded Peru in the 16th century, however, they discovered the luxuriant wool, and it soon became fashionable back home in Europe. The Spaniards also brought with them guns, enabling them to shoot vicuñas in large numbers. Skinning the dead animals made it possible to obtain much more of the precious wool in a shorter time. When Spanish rule ended in 1825, the vicuña was made a protected animal. Enforcement of the law was difficult, however, because the value of the wool remained high. Large numbers of vicuñas continued to be killed.

The huge demand for vicuña wool remained during the 19th and 20th centuries, and was often satisfied by shooting the animals. Killing tens of thousands of vicuñas every year ensured that they became steadily rarer, and by 1960 there were only about 6,000 left.

## Protection

Although the vicuña population has increased steadily since the 1960s, the animal has disappeared from large parts of its former range, and the remaining herds are widely scattered. More than half the vicuña population now lives in the Pampas Galeras National Park in Peru.

Today the vicuña is fully protected, and international trade in its wool is strictly controlled. Only wool shorn from living animals that can then be released can be legally sold. Such measures should ensure that further losses do not occur and that the animal will not become extinct.

**The vicuña** *is a member of the camel family and resembles a llama. The fine wool coat of the animal has evolved as protection against conditions of extreme cold high in the Andes Mountains.*

EX
EW
CR
EN
VU
NT
LC
O

# Gray Whale

## *Eschrichtius robustus*

*One of the most charismatic and best known of all cetaceans (whales and dolphins), the gray whale was one of the first of its kind to receive real protection from whalers. Today the species faces other problems, namely, increasing coastal disturbance and pollution.*

Gray whales may have done more to raise the profile of whales than any other large species. This is mainly because of their coastal habitat; they feed, breed, and migrate in shallow water, and so can be watched from the shore. There is much for spectators to admire, for grays are one of the most active whale species, regularly indulging in antics such as breaching (leaping out of the water), lobtailing (smacking the water's surface noisily with the tail), and spy-hopping (bobbing upright with the head out of the water to look around). The whales seem to relish playing in the surf off gently sloping coasts and frequently beach for hours, only to refloat on the next tide, to the relief of concerned onlookers. Grays are widely recognized as among the friendliest whales as well as the easiest to watch. They are curious and approach boatloads of whale-watchers so readily that at times it is hard to know who is watching whom!

However, relations between people and gray whales have not always been so good. In whaling circles the gray earned the nickname of "devilfish" because mothers would often charge at whaling boats in desperate attempts to protect their calves. The gray whale was in fact one of the first large whales to be targeted by human hunters, since its liking for shallow inshore waters meant that it could be easily trapped and killed. There is evidence that aboriginal peoples on both sides of the North Pacific were successfully hunting them as early as 2,000 years ago.

### Slow Swimmers

The gray whale also has the disadvantage of being a slow swimmer, which meant that in the 17th and 18th centuries it could be hunted from sailing ships. Whalers from Norway, Britain, and the United States hunted the gray whales of the North Atlantic to extinction toward the end of the 18th century. A century later the breeding populations on both sides of the Pacific seemed to have disappeared as well.

### DATA PANEL

**Gray whale (California gray whale, devilfish, scrag whale)**

***Eschrichtius robustus***

**Family:** Eschrichtidae

**World population:** 15,000–22,000

**Distribution:** Eastern and western regions of the North Pacific

**Habitat:** Shallow coastal waters

**Size:** Length: 40–50 ft (12–15 m); females larger than males. Weight: 17–39 tons (15–35 tonnes)

**Form:** Robust-bodied animal with relatively small, narrow head; skin mottled dark gray, often with extensive patches of encrusting barnacles and algae. There is no dorsal fin, but a row of bumpy "knuckles" extends from the hump to the large tail fluke; throat has just 2 or 3 furrows

**Diet:** Bottom-dwelling organisms, especially crustaceans

**Breeding:** Single young born in warm waters in winter after 13-month gestation; weaned at 7 months; mature at about 8 years. May live up to 70 years

**Related endangered species:** No close relatives, but many other whales are threatened

**Status:** IUCN LC

By the time deep-sea whaling got underway in the late 19th century, the gray whale had already become rare, and attention had shifted to the more lucrative blue and fin whales. Fewer than 1,200 gray whales were caught between 1921 and 1947, at which time an international ban on commercial hunting of this species was put in place. Thanks largely to the ban, the eastern Pacific population of gray whales now numbers more than 15,000 individuals, and their recovery has been complete enough to allow the resumption of small-scale whaling by such aboriginal peoples as the Inuit and the Makah tribe of Washington State.

## Modern Threats

So-called "traditional" hunts are still controversial, but probably less damaging than other more modern threats to the gray whale population. For example, many whales die every year after becoming entangled in fishing nets, and grays are also affected by chemical pollution, coastal development, and noise. The busy shipping lanes of the West Coast of the United States mean that on migration most gray whales now choose to travel up to 40 miles (65 km) offshore, while in earlier times the whole population often used to pass within 2 miles (3 km) of the coast.

On the other side of the Pacific there has been no such recovery, and a gray whale sighting close to Japanese shores is rare enough to be newsworthy. The remaining western Pacific population probably numbers only a few hundred individuals. As a whole, the IUCN has downgraded the species from Endangered to Least Concern, but that status is dependent on strict hunting controls remaining in place.

**Gray whales** *can reach a length of 50 feet (15 m). Their skin is mottled and heavily parasitized; barnacles and whale lice live on it. Young grays are smooth in comparison to the elders, as this picture of a mother and her calf shows.*

EX
EW
CR
EN
VU
NT
LC
O

# White Whale

## *Delphinapterus leucas*

*The white whale, or beluga, exhibits a diverse array of unusual and engaging characteristics. Sadly, however, its numbers have declined because of a long history of exploitation.*

Belugas are among the most vocal whales. Many of their more musical calls can be heard above the ocean's surface, earning the species the nickname "sea canaries." Below the waves, however, the clicks, trills, moos, whistles, squeaks, and twitters produced by a herd of beluga has been likened to the noise of a busy farmyard! Although some of these sounds are undoubtedly used for communication, many are involved in echolocation; the whales have a distinctive bulge—known as the melon—on their foreheads, which they use for focusing pulses of sound on underwater obstacles or prey.

Belugas can survive in extremely shallow water, an ability they put to good use when hunting. Herds of belugas have been known to drive schools of fish into shallow bays or up sloping beaches until they have nowhere to escape. Their taste for fish, which sometimes takes them hundreds of miles up rivers, brings them into competition with fishermen, and at one time the whales were killed to preserve fish stocks. Today less drastic methods may be used to scare the animals away, such as playing them recordings of predatory killer whales.

White whale meat is used for both human and animal consumption, especially by aboriginal peoples of the Arctic and subarctic regions. The whale's oily blubber, which can be up to 8 inches (20 cm) thick, serves to make soap, margarine, and various lubricants; the oil extracted from the head is considered to be of especially fine quality. Beluga bones are ground up for fertilizer, and the hide is tanned to make soft leather. Commercial hunting is now restricted by the International Whaling Commission, but natives of Greenland and North America are

### DATA PANEL

**White whale (beluga)**

***Delphinapterus leucas***

**Family:** Monodontidae

**World population:** More than 150,000

**Distribution:** Arctic Ocean and some adjoining seas

**Habitat:** Fjords, estuaries, and other shallow coastal waters of Arctic and cold temperate seas

**Size:** Length: 10–15 ft (3–4.6 m); males larger than females. Weight: 1,100–3,300 lb (500–1,500 kg)

**Form:** Chubby, dolphinlike animal; large, bulging forehead, small eyes, expressive face; distinct, flexible neck; no dorsal fin; adults have unmistakable white skin

**Diet:** Fish, including cod, salmon, and herring; also octopus, squid, crabs, and various bottom-dwelling crustaceans, mollusks, and worms

**Breeding:** Single young born every 2–3 years in April–September after 14-month gestation; weaned at 18–24 months; females mature at 4–7 years, males at 8–9 years. May live up to 35 years

**Related endangered species:** Narwhal (*Monodon monoceros*) NT

**Status:** IUCN NT

**Belugas** *are born with dark-brown or black skin (left) that gradually fades to white as the whale matures.*

still permitted to kill some animals every year. Hunting white whales is easy because they come so close to shore. The number of animals killed in 2006 was 456 but it varies from year to year.

### The Threat from Hunting and Pollution

Some of the more accessible beluga populations have already been hunted to extinction, and others have declined severely. For example, the resident population in Canada's Gulf of St. Lawrence, which once numbered over 20,000 animals, was reduced by the 1970s to just 300. For a long while it seemed that even with legal protection the population would never recover, since the waters of the gulf had become so polluted that the belugas often failed to breed. Of the few calves that were born, many had serious deformities that made them unable to swim. At one time levels of pollutant in white whale tissue were so high that the corpses of whales were themselves disposed of as toxic waste. It seems, however, that the story may have a happy ending. Strict hunting and pollution controls have begun to take effect, and the population is starting to recover; by 2004 there were 900–1,000 white whales in the Gulf of St. Lawrence.

# Trumpeter Swan

## *Cygnus buccinator*

*The largest of the native North American wildfowl and the biggest of the world's seven species of swan, the magnificent trumpeter came close to extinction through hunting.*

The trumpeter swan is named after its loud, deep, trumpeting calls. Largely as a result of hunting, it is the rarest of all the world's swans. In the past the trumpeter swan nested from Alaska across much of Canada and into the United States as far east as Indiana and south to Missouri. However, from the European settlement of North America in the 17th century to the early years of the 20th century the trumpeter swan was hunted for its meat, skin, and feathers, and its eggs were taken, too. The meat and eggs were regarded as delicacies, while the tough skins, with their warm, soft, downy feathers, were made into boas, hats, and other articles of dress, and the down into powder puffs.

### In the Nick of Time

In 1918, as a result of alarm at the catastrophic decline of various birds, including the trumpeter swan, hunting was outlawed by the Migratory Bird Treaty Act. The act almost came too late. By 1932 only 69 trumpeter swans were known to exist, all of them in one area, near Yellowstone National Park. In 1935 the federal government designated about 22,000 acres (9,000 ha) of prime trumpeter nesting habitat in

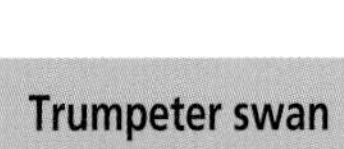

## DATA PANEL

**Trumpeter swan**

***Cygnus buccinator***

**Family:** Anatidae

**World population:** About 35,000 birds (2005 estimate)

**Distribution:** Major breeding areas in Alaska and northwestern Canada; isolated breeding populations elsewhere, including Michigan, Wisconsin, Minnesota, and Oregon. Some populations migrate: main wintering areas are on the coasts of Alaska, British Columbia, and Washington State

**Habitat:** Breeds by freshwater lakes, ponds, and marshes; winters on shallow lakes and reservoirs, and by streams and rivers; also on estuaries and sheltered coasts; forages in croplands and pastures

**Size:** Length: 4.9–5.1 ft (1.5–1.8 m); wingspan: 7.3–8.2 ft (2.2–2.5 m). Weight: 16–27.5 lb (7.3–12.5 kg)

**Form:** Large swan. Adult plumage entirely white, contrasting with black bill, short black legs, and black webbed feet. Heads and necks often stained a rusty color due to immersion in iron-rich waters when feeding. Juveniles have gray-brown plumage and a mainly pinkish bill

**Diet:** Plant food, especially sago pondweed and duck potato tubers; also stems, young shoots, leaves, seeds, roots, and tubers of other aquatic plants; grasses and grain crops in fields. Young are fed on aquatic invertebrates and fragments of vegetation

**Breeding:** Establishes mate for life when about 3 years old; breeds the following year. In late March to early May male gathers marsh plants and brings them to female, who builds a huge nest mound on a shore or islet or on a beaver or muskrat lodge. Female lays 3–9 large, white eggs which she incubates for about 5 weeks; her mate stands guard. Downy young (cygnets) grow quickly; by 8–10 weeks they have reached half their adult size and are fully feathered. Young fly at 3–4.5 months; stay with parents for their first winter and then return to the breeding areas

**Related endangered species:** None

**Status:** IUCN LC

southwestern Montana's Centennial Valley as the Red Rock Lakes National Wildlife Refuge. The Red Rock Lakes population served as a source of breeding stock for reintroductions of the species to parts of its original range, chiefly to wildlife refuges in the Midwest.

**The trumpeter swan** *has a wingspan of 8 feet (2.4 m) and a long neck, which it holds erect when swimming, not curved, as in the mute swan. The trumpeter's bill is blackish rather than orange and black.*

## Against the Odds

Trumpeter swans are still vulnerable to shooting, even in error: The much smaller tundra or whistling swan, and even the snow goose, can be mistaken for the trumpeter at long range. Trumpeters are also killed by lead poisoning as a result of swallowing lead shot in cartridges and sinkers used by anglers to keep their lines under water. Lead shot has been banned nationally for wildfowl hunting in the United States since 1991, but trumpeters still succumb, since old pellets can stay in the sediment at the bottom of lakes and wetlands for decades. Other threats to the swans include collision with power lines and wire fences that snake across the land. Pollution and disturbance by boats and even birdwatchers can cause problems. The swans' winter habitat is under threat from pollution. Another problem today is that some of the reintroduced flocks have lost most of their migratory instinct and move only short distances south to become crowded in unsuitable wintering areas where they may suffer freezing and starvation.

Despite such threats, populations of trumpeter swans have increased from about 2,000 birds in the early 1960s to about 35,000 in 2005, with several populations showing continued increases. Conservation steps have included reestablishment of swans in breeding haunts and wintering quarters, protection of their habitat, and public education programs.

EX
EW
CR
EN
VU
NT
LC
O

# Galápagos Cormorant

## *Phalacrocorax harrisi*

*Isolated for thousands of years by the remoteness of its native islands, the Galápagos cormorant lacks both an instinct for danger and an ability to fly. Its refuge has now been invaded by people, pollution, and alien predators, and its naturally small population has fluctuated dramatically in recent years.*

The Galápagos Islands are renowned for their unique wildlife, yet few of their native species are more spectacularly odd than the flightless Galápagos cormorant. It lives only on Fernandina and the north and west shores of Isabela, on the western side of the equatorial Galápagos. The shores are bathed by the cool, nutrient-rich waters of both the Cromwell and Humboldt Currents. The nutrients support a lush growth of plankton that feeds vast shoals of small fish, so for much of the year the seas around the cormorant colonies are teeming with food.

For the Galápagos cormorant the land is just somewhere to breed; it feeds in inshore waters. Like all cormorants, it hunts under water for fish, diving beneath the surface and driving itself along with its webbed feet. Wings just get in the way while fishing, so over the millennia those of the Galápagos cormorant have been reduced to threadbare stumps. Its wing muscles have dwindled too, along with the deep keelbone to which they were attached. All the energy has been diverted into its legs and feet, which are unusually big and strong. They give the cormorant the power to swim right down to the rocky seabed, flush a fish or octopus from a crevice, and pursue and catch it in its hooked bill before surfacing to eat it.

### DATA PANEL

**Galápagos cormorant**

***Phalacrocorax harrisi***

**Family:** Phalacrocoracidae

**World population:** 1,700 birds (2006 estimate)

**Distribution:** Coastlines of Fernandina and Isabela in the Galápagos Islands

**Habitat:** Breeds on ledges of volcanic rock just above shoreline; feeds in inshore waters

**Size:** Length: 35–39 in (89–100 cm); each wing averages 10 in (25 cm) in length. Weight: 5.5–8.8 lb (2.5–4 kg)

**Form**: Heaviest of all cormorants (males heavier than females). Powerful hooked bill and long, thick, sinuous neck; long, broad tail; strong, fully webbed feet; reduced weak wings with sparse plumage; blackish brown plumage with bare throat patch; pale green-blue eyes. Juveniles glossy black with brown eyes

**Diet:** Marine fish, mainly bottom-dwelling species such as eels and rockfish; also octopus and squid

**Breeding:** March–September. Nests in groups of up to 12 pairs, building bulky nests of seaweed on remote sites near shore. Up to 4, usually 2–3 whitish eggs incubated for 5 weeks by both sexes; young fledge at 8.5 weeks, but stay at nest for a further 4.5 weeks, where they are often fed by male alone

**Related endangered species:** Nine, including bank cormorant *(Phalacrocorax neglectus)* EN; Socotra cormorant *(P. nigrogularis)* VU

**Status:** IUCN VU

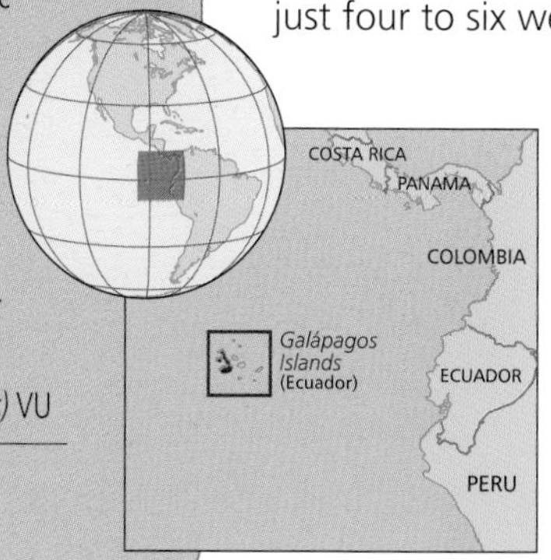

COSTA RICA
PANAMA
COLOMBIA
*Galápagos Islands* (Ecuador)
ECUADOR
PERU

### Climatic Hazard

Galápagos cormorants typically breed between the months of March and September, when the cool, rich ocean currents guarantee their food supply. However, in December the trade winds that drive the Humboldt Current retreat southward, allowing warmer, virtually sterile surface water to flow in from the north. This seasonal current is called El Niño. It usually lasts for just four to six weeks, but every four or five years an exceptionally powerful El Niño event disrupts the marine ecosystem for up to nine months. The plankton supply fails, the fish vanish, and the seabirds starve.

The cormorants have had to cope with regular setbacks since they arrived on the islands, and they show an amazing ability to recover. In 1982 a catastrophic El Niño event slashed their population from some 850 to a perilously low total of 400 adults, but 18 months later the population had more than doubled to about 1,000 birds. There was another unusually extreme El Niño event in 1997, and another in 2009. Some researchers fear that global climate change may increase the frequency of such events.

### Enemy Aliens

Even at their peak, Galápagos cormorants have never been numerous. They are also restricted to just 230 miles (370 km) of coastline on their native islands. This makes them vulnerable to any local disaster such as the oil spill caused by the grounding of the oil tanker *Jessica* on San Cristóbal Island in January 2001. San Cristóbal is on the eastern side of the Galapágos, but the winds and currents drove the oil slicks west, threatening the southern and eastern shores of Isabela. The cormorants had a narrow escape, since they inhabit the northern and western shores.

Oil is not their only problem. Diving cormorants regularly drown in stray fishing nets and lobster traps. Commercial fishing is illegal, but the waters are so rich in tuna and other valuable species that they attract operators prepared to flout the rules.

Fishermen may also import alien predators such as rats, cats, and even wild dogs. Lacking both the instinct to escape danger and the ability to fly, the cormorants, their eggs, and their young are vulnerable to such threats. Another problem is tourism. Although it helps fund the conservation of the Galápagos, tourism causes disturbance and also adds to the pollution problem; much of the oil spilled by *Jessica* was a delivery of fuel for tourist vessels.

**The Galápagos cormorant** *has little use for its wings, the muscles of which are weak. When fishing, the bird employs its strong legs and feet.*

# Harpy Eagle

## *Harpia harpyja*

*The world's most powerful bird of prey, the magnificent harpy eagle—like so many other large birds of the New World tropics—has suffered badly from hunting and is becoming increasingly disturbed by human encroachment into its rainforest habitat.*

The harpy eagle was named for the half-bird, half-woman creatures of ancient Greek mythology, the harpies, who were capable of snatching living things and bearing them out of this world; the name means "one who snatches."

True to its name, the harpy eagle is a swift and powerful hunter. With its incredibly sharp eyesight and hearing, it can locate prey at a considerable distance. The harpy's main targets are sloths: The slow-moving mammals are probably particularly susceptible because they are conspicuous when they climb up into the canopy (rooflike covering of trees) to warm themselves in the morning sun.

The eagle begins its attack by swooping down from its perch high in the trees at impressive speed, weaving neatly between the branches with surprising agility for such a huge bird, thanks to its short, broad, and rounded wings and its long tail. Then, as it closes in on its victim, it suddenly rolls over on its back and, as it hurtles past beneath the sloth, rips the sloth from the branch, unlocking the grip of the mammal's huge claws. Righting itself immediately, the eagle then flies back to the canopy, where it tears up the prey to feed itself or its young. To achieve such a feat, the harpy eagle has the mightiest legs and feet of any bird of prey. Its legs are as thick as a child's wrist and capable of carrying prey almost as heavy as the bird itself. Its sturdy toes are armed with huge, razor-sharp talons.

### Human Threats

The main threat facing the harpy eagle is shooting by hunters. Although the birds are generally inconspicuous and spend much of their time high in the trees, their great size and boldness when faced with humans, especially near the nest, makes them highly vulnerable to hunters. It is also likely that the eagles suffer food shortages when hunters kill animals on which they depend for food.

The threat from hunting is particularly acute because harpy eagles are extremely slow breeders. Although the female lays two eggs, one is wasted: As soon as the first chick hatches, she stops incubating the other egg. Also, because each youngster remains dependent on its parents for food and protection for up to a year or more after it has left the nest, a pair of harpy eagles can rear only one offspring at best every third year.

Although harpy eagles are still fairly common in Guyana and the great Amazonian forests, especially in Brazil and Peru, they have been exterminated from large areas of their former range. Some may be able to survive in disturbed areas of forest and even mixed forest and pasture, but harpy eagles are also threatened in the long term by the continued destruction or degradation of their rainforest habitat as a result of logging, mining, agriculture, and other developments. The situation is particularly acute in Central America, where the forests are generally smaller and more fragmented. With habitat alteration come roads, which make it so much easier for people, including hunters, to reach previously remote areas where the eagles were once safe.

### Conservation

Conservation plans to save the harpy eagle include surveying breeding birds and monitoring nests

**The harpy eagle** *has a powerful hooked bill with which it tears the flesh of its prey after killing it with its 2.8-inch (7 cm) rear talons.*

in parts of their range—particularly in Panama and Venezuela—to learn more about their status, distribution, and breeding biology. Satellite transmitters and radios have been attached to young harpy eagles so that their movements can be tracked. The data collected will help show how large a territory each bird needs to survive.

A captive-breeding program at the Peregrine Fund's World Center for Birds of Prey in Boise, Idaho, succeeded in rearing 10 young harpy eagles in the 1990s, which were later released in Panama, where suitable habitat remains. Another captive-breeding center operates in Panama itself, at the Neotropical Raptor Center, not far from Panama City. In the long term, however, it will be essential to control hunting much more effectively and to reduce the rate of forest destruction by setting up fully protected forest reserves.

## DATA PANEL

**Harpy eagle**

***Harpia harpyja***

**Family:** Accipitridae

**World population:** Estimated at 20,000–50,000 birds

**Distribution:** Mexico to northeastern Argentina

**Habitat:** Mainly large, undisturbed areas of lowland tropical forest

**Size:** Length: 3–3.5 ft (0.9–1 m); wingspan: up to 6.5 ft (2 m). Weight: male 9–10.5 lb (4–4.8 kg); female 16.8–20 lb (7.6–9 kg)

**Form:** Large bird of prey with relatively short, broad wings and long tail; head has double crest of stiff feathers; massive, sharply hooked bill; powerful legs and feet. Gray, black, and white plumage. Immatures take 4 years to reach adult plumage; they are duller in coloration with white head

**Diet:** Mainly tree-dwelling mammals such as sloths; also howler and capuchin monkeys, porcupines, and anteaters. Sometimes such ground-dwelling mammals as agoutis and young brocket deer; birds such as macaws and curassows; reptiles such as snakes and iguanas

**Breeding:** Builds huge nest up to 5 ft (1.5 m) wide in tall forest trees above the canopy; female lays 2 white eggs, incubated for about 8 weeks; only 1 young survives; fledging takes place at about 5 months

**Related endangered species:** Philippine eagle *(Pithecophaga jefferyi)* CR; New Guinea harpy eagle *(Harpyopsis novaeguineae)* VU

**Status:** IUNC NT

EX
EW
CR
EN
VU
NT
LC
O

# Piping Plover

## *Charadrius melodus*

*The piping plover is an energetic little shorebird. It has been threatened by human pressure on its habitat and has declined considerably since the 1950s. However, a huge effort is being made to help the bird, and recent increases have resulted in it being downlisted.*

The piping plover is the rarest of the ringed and sand plovers that breed in North America. It has not fared well in the modern world. Its troubles began when hunters shot large numbers during the 19th century. From about 1850 its range contracted considerably, and by 1900 the toll taken by hunting had brought the piping plover to the verge of extinction. In the nick of time changes were introduced to the game laws, and protection allowed the species to recover considerably by the 1920s.

In more recent times the piping plover has suffered another decline, this time as a result of disturbance to its habitat by humans. The building of roads, houses, and other developments in coastal areas have all had an adverse effect on numbers.

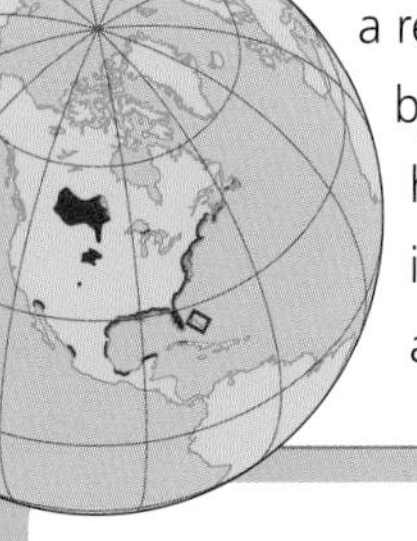

As soon as the birds arrive on the breeding sites from their wintering grounds, human activity can interfere with the birds' ability to establish territories and their courtship. Because their eggs and young are so well camouflaged, they are easily trampled or dispersed by people visiting beaches; and if disturbance is severe, the birds may abandon attempts to breed in the area. The use of off-road vehicles on breeding beaches destroys plover nests. Pet dogs allowed to run free eat eggs, kill chicks, and frighten adults, while garbage and food scraps dumped on beaches attract other predators such as foxes, rats, possums, skunks, feral cats and dogs, and gulls.

In 1985 the piping plover was placed on the Federal list of Threatened and Endangered Species. The Great Lakes population received Endangered status, while the Great Plains and East Coast populations

### DATA PANEL

**Piping plover**

***Charadrius melodus***

**Family:** Charadriidae

**World population:** 12,000–13,000 birds

**Distribution:** Breeds on northern Great Plains of Canada and U.S. and along Atlantic coast from Newfoundland to North (and occasionally South) Carolina; also around Great Lakes. Winters on Atlantic coasts of southern U.S.; also Bahamas, Gulf of Mexico, northwestern Mexico, and eastern Caribbean

**Habitat:** Great Plains breeders nest on sand and gravel shores of large, alkaline lakes; also on river banks, sand and salt flats, and floodplains. Coastal breeders nest on sandy beaches and winter at long-used sites on sandy beaches and sand flats; sometimes on estuary mudflats and dunes

**Size:** Length: 6.8–7 in (17–18 cm); wingspan: 14–15 in (35–38 cm). Weight: 1.5–2.3 oz (43–64 g)

**Form:** Small, plump-breasted bird with orange legs; short, black bill (orange with black tip in breeding season); pale sandy-gray upperparts, white underparts; black bar across forehead and black "collar" in breeding season (both pale in winter). White rump, visible in flight, distinguishes it from more common semipalmated and snowy plovers

**Diet:** Aquatic invertebrates, including worms, crustaceans, mollusks, and insects; also land-dwelling insects such as flies, midges, beetles, and grasshoppers

**Breeding:** Female lays 3–4 dark-spotted buff eggs in April or May; incubated by both parents for about 4 weeks; chicks cared for by both parents; fledge after about 4 weeks

**Related endangered species:** Mountain plover *(Charadrius montanus)* NT; New Zealand dotterel *(C. obscurus)* EN; St. Helena plover *(C. sanctaehelenaei)* CR; hooded plover *(C. rubricollis)* VU

**Status**: IUCN NT

**The piping plover** *breeds in the United States and Canada and has a winter range that extends from North Carolina south to Jamaica and northern Mexico.*

were listed as Threatened. A recovery plan was approved in 1988. As a result of various conservation measures, numbers are now stable or increasing, particularly along parts of the Atlantic coast, and especially in New England.

International censuses of the piping plover involving eight countries are carried out every five years for wintering and breeding populations. The breeding census shows that the Atlantic coast populations are faring much better than those in the interior of the United States and Canada. In the northern Great Plains breeding numbers have decreased every year as a result of drought and flooding, often caused by water-management projects. In the Great Lakes area there is a viable, though still very small, breeding population in Michigan.

The Great Lakes birds may suffer from high levels of toxic pollutants, especially PCBs (polychlorinated biphenols). Piping plovers, especially males, are extremely loyal to breeding sites, returning year after year, so it is a serious loss if a particular site is damaged or destroyed.

Since the piping plover spends as much as 70 to 80 percent of its annual cycle away from its breeding range, it is vital to protect its wintering habitat, too. Conservationists are focusing attention on the Laguna Madre region of Texas and Mexico, where there are vast areas of winter habitat. Such wilderness, also susceptible to development, could save many piping plovers and other endangered birds. Other threats that need constant vigilance are oil spills and dredging operations in the Gulf of Mexico.

# Red Siskin

## *Carduelis cucullata*

*The strikingly attractive red siskin has been brought to the verge of extinction chiefly as a result of the relentless demands of the cage-bird trade.*

Until the early years of the 20th century the red siskin was common in northernmost South America. Once it could be found throughout the foothills of the mountain ranges of northern Venezuela at altitudes of between 900 and 4,250 feet (280 and 1,300 m); it was also found in northern Colombia. There was a very small population on the island of Trinidad, which originated either from birds taken there and deliberately introduced into the wild, or from cage birds that had escaped from captivity. The species was also introduced to Puerto Rico during the 1920s or 1930s, and the small numbers recorded in the wild in Cuba suggest that these birds had escaped from captivity on the island.

Today, by contrast, the red siskin is extremely rare and is restricted to a few fragments of its former range. While in the past it could be found in 15 different states in Venezuela, in recent years there have been only a few sightings from just four states. The only known population in Colombia is a small one in the state of Norte de Santander on the northeastern border with Venezuela. The Trinidad population has probably disappeared entirely, and there are very few recent records of the species from Puerto Rico. However, in 2000 a new population was discovered in southwest Guyana.

### Ruthless Trapping

The main reason for this little finch's catastrophic decline has been trapping for the cage-bird trade. Since the 1940s trade in the birds has been illegal, but trappers—who supplied a huge demand for red siskins—were not to be deterred. Trapping continued on a massive scale for much of the 20th century.

The demand was not so much for the intrinsic beauty of the birds themselves as for their use in interbreeding with domestic canaries. Breeders found that the red siskins were closely enough related to the domestic birds to attempt to introduce genes for red color into the domestic stock, thus producing the highly sought-after "red-factor canaries."

However, because domestic canaries belong to a different genus from that of the red siskin, breeders have found it difficult to transfer the "red" gene into canary stocks. Although they have bred types they call "red canaries," these

### DATA PANEL

**Red siskin**

***Carduelis cucullata***

**Family:** Fringillidae

**World population:** 2,500–10,000

**Distribution:** Recent sightings in 4 states in northern Venezuela and in NE Colombia; used to occur much more widely. In 2000 a population was found in SW Guyana. There were once populations of escaped cage birds in Puerto Rico and on Trinidad

**Habitat:** Moves seasonally and daily between moist evergreen forest, dry deciduous woodland, and adjacent shrubby grassland and pastureland

**Size:** Length: 4 in (10 cm)

**Form:** Small, brightly patterned red-and-black finch. Male has jet-black head, chin, throat, and tail; also black wings with broad red bar across flight feathers; rest of plumage rich scarlet to pinkish red. Female is brownish gray from head to back; crown, nape, and "shoulders" streaked darker; rump and wingbars bright orange-red

**Diet:** Seeds of trees and shrubs; grass seeds, flower heads, and cactus fruit

**Breeding:** Main nesting season May–early June; second breeding period in November and December. Builds neat, cup-shaped nest in tall trees

**Related endangered species:** Yellow-faced siskin *(Carduelis yarrellii)* VU; saffron siskin *(C. siemiradzkii)* VU; Warsangli linnet *(C. johannis)* EN

**Status:** IUCN EN

birds have a reddish tinge rather than the splendid bright coloration of the red siskin.

In more recent years keepers of cage birds have shown more interest in the species for its own undeniable beauty. However, the trapping of wild birds did not stop, despite the species' rarity. In 1975 almost 3,000 red siskins were known to have been caught. Even in 1981, when total numbers were doubtless far fewer, just over 1,000 birds were still trapped. After breeding, the birds would gather in large flocks, which were easily lured to baits and caught; but recently only single birds or pairs have been seen. As if this relentless trapping was not enough of a threat, red siskins also face loss of habitat as a result of the spread of intensive agriculture.

**The red siskin** *has long been a target for cage-bird traders. Large numbers went to breeders intent on introducing the genes for red color—particularly prominent in the male (above)—into domestic canaries.*

## A Difficult Task

The task that lies ahead for conservationists trying to save the red siskin is not an easy one. Despite being legally protected in Venezuela and being listed on CITES Appendix I (see page 60), trapping may still continue as a result of the difficulties of enforcing the law on the ground. There have been no records of the species for many years from the Venezuelan national parks of Guatopo and Terepaima—the only protected areas where the birds were reputed to have occurred.

**A female red siskin** *(below) does not have the bright plumage of a male.*

Although captive-breeding programs have been initiated, they have been beset by problems of disease and genetically impure stock; a project to reintroduce the species to Trinidad had to be abandoned because of disease. Although there is a need to raise public awareness of the species' plight, some past education initiatives have had the opposite effect, leading to an increase in pressure from the cage-bird trade. No one knows how many red siskins are left in the wild. Recent estimates suggest just a few thousand, and it may not be much longer before it is impossible to save the species in the wild.

# Alabama Red-bellied Turtle

## *Pseudemys alabamensis*

*In 1987 the Alabama red-bellied turtle was designated an endangered species by the United States Fish and Wildlife Service. Its reliance on clean water and sandy beaches has made it vulnerable to habitat destruction.*

The distribution of the Alabama red-bellied turtle is limited to quiet backwaters, pools, tributaries, and rivers in the Mobile Delta in Mobile and Baldwin Counties. The turtle seems to prefer shallow- to medium-depth water with a good growth of aquatic plants and a layer of silt into which it can dive if disturbed. Today red-bellies are found in scattered areas, although they once had a wider range. Urban development, drainage, and other human activities have restricted the turtles' territory.

Red-bellies are semiaquatic (adapted to a life both on land and in water). They spend much time basking in the sun to maintain a suitable body temperature and enter the water to forage on aquatic plants.

### DATA PANEL

**Alabama red-bellied turtle**

***Pseudemys alabamensis***

**Family:** Emydidae

**World population:** Unknown

**Distribution:** Streams running into Mobile Delta, Alabama

**Habitat:** Freshwater streams; rivers with muddy bottoms

**Size:** Length: females 13 in (33.5 cm); males 11 in (29.5 cm)

**Form:** Shell with yellow and black ocelli (eye spots) on scutes (shields). Carapace (upper shell) green; plastron (lower shell) orange; colors darken with age. Head and limbs brown with yellow stripes

**Diet:** Aquatic plants; captive specimens take fish, meat, and earthworms, suggesting that in the wild they also feed on aquatic vertebrates

**Breeding:** One clutch per year of 3–9 eggs

**Related endangered species:** Rio Grande cooter *(Pseudemys gorzugi)* NT; American red-bellied turtle *(P. rubriventris)* NT

**Status:** IUCN EN

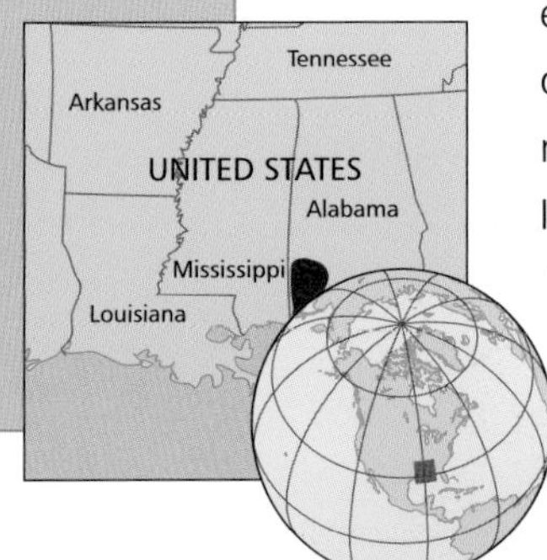


The species has not been thoroughly studied, and knowledge of various aspects of its lifestyle is still limited. Even its life span is unknown, although like many other chelonian species (turtles and tortoises of the order Chelonia) it may be able to live for 50 years or more. The number of eggs laid is also unknown; if it is at the lower end of the estimated range—three to nine per year—then recovery of the species could be slow and difficult. However, one reason why turtles have been an evolutionary success is their longevity rather than their rapid breeding rates.

### At the Mercy of Humans

Red-bellied turtles have many natural predators such as alligators, fish crows, raccoons, and large fish. The turtles are also at the mercy of other creatures; crows and pigs destroy nest sites, and fire ants (an introduced pest) have been found attacking the eggs in turtle nests. However, humans have had by far the greatest effect on red-bellied turtle numbers. Adult turtles and their eggs used to feature in the human diet. During the Great Depression of the 1930s turtles of various species in many areas were eaten as an essential rather than a luxury item.

Habitat destruction is perhaps the greatest threat. Waterside sites with loose, sandy soil are essential for egg laying. Many such areas have been degraded by off-road vehicles. The major nesting site for red-bellies is Gravine Island in Baldwin County. Even in the 1980s, when the species was documented as being in decline, eggs were still taken by humans.

**The adult Alabama red-bellied turtle** *has a brownish carapace (upper shell covering the back) and a red plastron (lower shell covering the belly).*

In spite of legal protection, the red-belly is vulnerable to a number of factors. Pollution and pesticides are two potential threats. Lowering water levels and the introduction of plants that choke the waterways cause further habitat destruction.

## Collective Guilt

A substantial factor in the decline of the red-bellied turtle has been collecting by hobbyists and others. As is the case with many turtles, the hatchlings are appealing creatures and are often taken as pets. Hatchling red-bellies are particularly attractive, having a green carapace (upper shell) with dark-edged lines and yellow and black ocelli (eye spots) on the scutes (shields). The plastron (lower shell) is red to orange with varying dark markings. The head and limbs are olive to dark brown with yellow stripes. The colors darken with age—the carapace turns brown with reddish-yellow areas. The plastron also often loses the dark markings.

The fate of collected red-bellies has often been death because of ignorance of their requirements or simply neglect when the owner's interest has waned. Turtles need the correct diet, clean water, and sunlight (real or artificial) in order to thrive. Experienced keepers may go on to breed their charges, but all too often turtles that start off with a shell length of about 1 inch (2.5 cm) gradually outgrow their housing and are dumped in the nearest water.

Commercial traders used to take substantial numbers of red-bellied turtles to sell as pets. No doubt the practice still happens today to some extent. However, under Alabama nongame species regulations it is illegal to take, capture, kill, possess, sell, or trade red-bellies and various other reptiles, amphibians, and other animals without a scientific permit or written permission from the Department of Conservation and Natural Resources. A federal ban on the sale of all turtle species with a shell length of 4 inches (10 cm) or less was introduced in 1975 by the United States Food and Drug Administration. The ban has not been rigidly enforced and is often ignored, nor did it apply to the export of live turtles.

# Bog Turtle

## *Glyptemys muhlenbergii*

*The bog turtle is one of the smallest turtles in the world. Its colorful markings have made it a popular pet and led to overcollection. The species is also at risk from loss of habitat.*

The bog turtle, sometimes referred to as Muhlenberg's turtle, is probably the smallest turtle in the United States and one of the smallest in the world. Bog turtles take between five and seven years to reach maturity, by which time they are about 3 inches (8 cm) long. Eggs are not buried deeply but simply deposited in a small hole scooped out from loose soil or moss at the base of grass tussocks. Although it would seem to have a wide distribution—from Massachusetts down to South Carolina—there are only two major populations (north and south) divided by a substantial gap. Turtle groups in these areas are often widely scattered, and some sites may contain fewer than 20 specimens.

### Overcollection

Bog turtles are relatively colorful. They have light patches on the head, forelimbs, and throat, and their scutes (hornlike shields that form the upper shell) have central light-reddish to orange markings that sometimes resemble a "sunburst" pattern. The combined appeal of their small size and attractive appearance has partly been the cause of the bog turtle's decline. They have long been a target for collectors, who keep them as pets or sell them to the pet trade. This has continued even though in some states the species has been protected for over 20 years. However, the laws were seldom enforced, and collection continued unchecked. In 1997 the species was given federal listing as Threatened. Under this legislation anyone convicted of capturing, harming, or even transporting a federally threatened species could be fined up to $50,000 and imprisoned for up to one year. It is unlikely that this will deter all collectors, since the black market price for bog turtles is high.

Although it was decided not to designate Critical Habitat for the species, its protected status meant that construction projects could be stopped or delayed if turtles were found to be present on the site. There were even reports of bog turtles being deliberately planted to prevent work starting.

UNITED STATES

### DATA PANEL

**Bog turtle**

***Glyptemys muhlenbergii***

**Family:** Emydidae

**World population:** Unknown

Distribution: Eastern U.S.; scattered locations in New York state, New Jersey, Pennsylvania, Georgia, Maryland, Connecticut, Massachusetts, and the Carolinas

**Habitat:** Wetlands; slow-moving streams

**Size:** Length: 3–3.5 in (7.6–8.9 cm)

**Form:** Shell brown to black; forelimbs and throat spotted or striped; yellow or orange patch on head

**Diet:** Earthworms, insects, tadpoles, and slugs

**Breeding:** Usually 2–4 eggs (up to 6 possible)

**Related endangered species:** wood turtle *(Glyptemys insculpta)* EN; Pacific pond turtle *(Actinemys marmorata)* VU; spotted turtle *(Clemmys guttata)* EN

**Status:** IUCN CR

## Disappearing Habitat

Bog turtles have also declined due to habitat loss and fragmentation. Some of this loss is natural; bogs, ponds, and similar wetland areas are subject to a natural process known as "succession," when dead vegetation accumulates in the water, collecting and binding sediment and eventually becoming land.

However, the majority of the bog turtle's habitat has been destroyed by human activity, particularly the drainage and planting up of wet areas. Succession can be speeded up if plants gain a foothold. Many plant species grow rapidly in marshy ground, soon making turtle movement difficult and blocking the open areas they need for basking. Lowering the water level can also speed up succession. Construction of walls, roads, drainage channels, and other barriers fragments the habitat, preventing turtle movement and isolating them in small groups. The smaller the groups, the greater the adverse effects of inbreeding damage the long-term survival of the species. Further hazards are water pollution from fertilizers—which also accelerate plant growth—and toxic pesticides.

The number of turtles surviving in the wild is not known. They are secretive, and in their natural habitat they are well camouflaged, making population counts difficult. In some areas their presence went undetected until the 1980s. Several American zoos have captive-breeding groups, and substantial numbers are in private hands. Actual figures of captive bog turtles are impossible to estimate since many private owners keep quiet about their specimens, which may have been acquired illegally.

## Future Hopes

Bog turtles will breed in captivity, and clutches of up to six eggs—sometimes more than one clutch per year—have been recorded. In some collections several generations have been bred. Captive breeding may well be the answer to the bog turtle's decline. Provided that existing habitats are preserved or new ones constructed, youngsters could be released back into the wild. However, without this protection the decline of the bog turtle will continue.

**The bog turtle's** *attractive markings have made the species highly collectable, even though the practice is illegal.*

EX
EW
CR
EN
VU
NT
LC
O

# Desert Tortoise

## *Gopherus agassizii*

*Natural predators have always threatened the desert tortoise, but in the past few decades the greatest threat has come from humans.*

The desert tortoise has inhabited arid desert regions in the southwestern United States for over 10,000 years. It has now been adopted as the "state reptile" of California, although it is also found in parts of Utah, Arizona, and Nevada. The desert tortoise lives in areas with fewer than 5 inches (12 cm) of rainfall per year, summer temperatures of over 140°F (60°C), and below freezing in winter. Moisture from its food, and surface water when available, is stored in its bladder and will sustain the tortoise for many months. Desert tortoises avoid extremes of heat and cold by sheltering in burrows, often up to 30 feet (10 m) long, which they dig using their powerful forelimbs. During their active periods they scoop out shallow scrapes under bushes or rocky overhangs to use as shelter.

The desert tortoise and the related gopher tortoise are described as "keystone species" because many other animals depend on them. Their burrows are used by snakes, lizards, mammals, and various insects. If the tortoises and their habitat go, then these could also disappear. Natural predators such as coyotes, bobcats, foxes, skunks, ravens, and eagles have always preyed on tortoise eggs and young, but the greatest threat is now from collection and habitat destruction.

Tortoises are still popular as pets, but they are sometimes dumped in an unsuitable habitat when they are no longer wanted. The release of sick tortoises into the wild is thought to have caused severe outbreaks of upper respiratory tract disease in some wild populations. Once numbers are declining, simply maintaining a population level becomes difficult since breeding is a slow process. Females do not breed until they are between 15 and 20 years old; they may not breed at all in times of food shortage, and the survival rate of hatchlings is low. Due to harsh conditions and predators, only an estimated 1 to 3 percent survive to adulthood.

Habitat destruction is a major threat. Desert areas that were once of little use are now in increased demand for urban development, mining, waste, landfill, dumping, military training, and other uses. Increased road building and resulting traffic cause many tortoise deaths. Fire kills tortoises and destroys the vegetation that provides cover and food. Burned or otherwise damaged sites tend to be invaded by nonnative (alien) plants that are not the tortoises' preferred food. A similar

### DATA PANEL

**Desert tortoise**

***Gopherus agassizii***

**Family:** Testudinidae

**World population:** Unknown

**Distribution:** Sonoran and Mojave deserts

**Habitat:** Hot, arid deserts with firm sand, low growth, and open spaces

**Size:** Length: up to 15 in (38 cm); males are larger than females and can weigh almost 15 lb (7 kg)

**Form:** Domed, brownish shell with yellowish areas; slightly flattened fore-limbs; sturdy hind limbs. An extended shield at the front of the shell is used to flip other males during combat

**Diet:** Annual wildflowers, grasses, cacti, and other shrubs

**Breeding:** Two or 3 clutches of 2–14 eggs per year

**Related endangered species:** Gopher tortoise *(Gopherus polyphemus)* VU; Bolson tortoise *(G. flavomarginatus)* VU

**Status:** IUCN VU

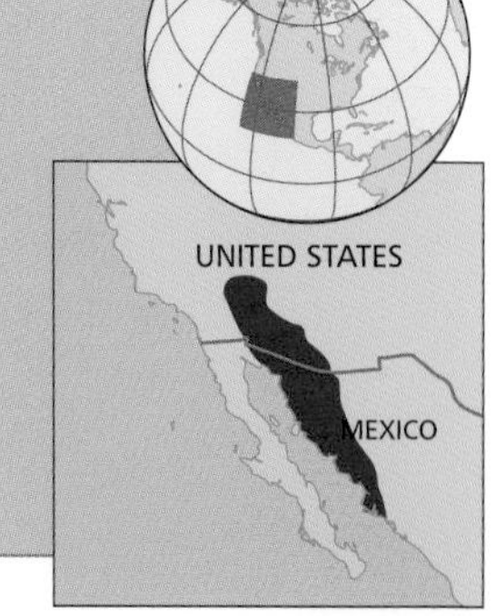

loss of vegetation and invasion by alien plants results from overgrazing by cows, sheep, and feral burros (small donkeys). These grazers also disturb the soil, compacting nests, and possibly trampling emerging or sheltering hatchlings. Off-road vehicles have a similar effect on these fragile desert ecosystems.

## Tortoise Protection

In 1974 the Desert Tortoise Preservation Committee was formed to promote the welfare of the desert tortoise. Several reserves have been designated, some of which have fenced roads and culverts under the highway to allow safe passage.

Numerous government agencies at federal, state, and local level are involved in habitat protection. The Federal Bureau of Land Management administers much of the reserved areas. State fish and game departments are also involved, while certain universities assist in ecological and population surveys.

**The desert tortoise** *has forelimbs specialized for excavation. Their burrows are often used by other animals, which form interdependent communities with the tortoises.*

The Federal Bureau of Land Management established a 38-square-mile (98 km$^2$) sanctuary in California called the Desert Tortoise National Area, in which livestock grazing and mining are banned, and which is closed to vehicles.

All desert tortoises are now protected by law; it is an offense to kill them, molest them, or sell or buy them, and permits are needed for captive specimens, which must be tagged. The permit system is organized by the California Turtle and Tortoise Club (CTTC). The only legal way to obtain a desert tortoise is by using the CTTC adoption scheme or by being given a captive-bred specimen.

# Galápagos Land Iguana

## *Conolophus subcristatus*

*Land iguanas were once abundant on the Galápagos Islands—a place renowned for its unusual animal life. Like nearly all the other reptiles found there, they do not live anywhere else. Increasing human and animal populations have been a threat, but conservation efforts are achieving success.*

The land iguana is found on six of the 19 islands that make up the Galápagos archipelago west of Ecuador. When he visited the islands in 1835, the famous English naturalist Charles Darwin commented that the huge numbers of land iguanas and their burrows left no room to pitch a tent. He also noticed their similarity to the green iguana of the mainland and the Galápagos marine iguana; the variation between these iguana species contributed to the formation of his ideas on evolution.

The land iguanas have adapted to the hot, dry climate by feeding largely on the fleshy prickly pear cactus, moving the spiny leaves around the mouth until the spines break off or breaking them off with their claws. The iguanas spend part of the day basking in the sunshine, but they later take shelter under bushes or in burrows. On some islands females travel considerable distances to find suitable nesting sites, often in the craters of dormant volcanoes.

### Disturbance

Today the land iguanas on Santiago Island, where Darwin observed such large numbers, are extinct. Their numbers have also declined on other islands.

The land iguanas, like the Galápagos tortoises, have suffered from habitat destruction as the human population of the islands has increased. Land for housing and cultivation has taken some of the habitat, but the animals brought by people have been extremely destructive. Cats, domestic or feral (wild), will eat young iguanas, and dogs eat even the adult lizards. Goats and donkeys compete with the iguanas for plant foods—the former can crop bare an entire area before moving on. Feral pigs enjoy iguana eggs, which they root up. Feral dogs were responsible for near wipeouts of two iguana populations some 30 years ago. During the 1930s 70 land iguanas were transferred from

COSTA RICA
PANAMA
Galápagos Islands (Ecuador)

## DATA PANEL

**Galápagos land iguana**

***Conolophus subcristatus***

**Family:** Iguanidae

**World population:** Unknown

**Distribution:** Galápagos Islands, Pacific Ocean

**Habitat:** Tropical scrub forests with open spaces for basking

**Size:** Length: 4 ft (1.2 m). Weight: 10–15 lb (4.5–6.8 kg)

**Form:** Heavy-bodied lizard with a long tail and powerful limbs. Grayish-green, yellowish, or brownish body with a crest of conical spines along the back

**Diet:** Prickly pear cactus and other plants

**Breeding:** Up to 25 eggs are laid in burrows dug by the females. Incubation period of up to 4 months

**Related endangered species:** Barrington Island iguana *(Conolophus pallidus)* VU

**Status:** IUCN VU

Baltra Island to North Seymour as an experiment. Those remaining on Baltra became extinct when the island became a military base during World War II.

## Success Story

The Charles Darwin Research Station was opened in 1964 with the aim of saving the unique flora and fauna of the Galápagos Islands. Much of the early work was directed at saving the giant tortoises. In 1968 the Galápagos National Park service was set up to conserve wildlife. The two bodies cooperate in research and captive breeding. Breeding centers for tortoises on Santa Cruz Island and Isabela Island were successfully breeding tortoises when in 1980 a pair of land iguanas was installed in the Santa Cruz center.

By 1991 captive-bred iguanas were being released on Baltra after an absence of 50 years. Eradication of goats, wild dogs, and other pests was necessary and continues today. Cats and pigs still exist on some of the islands, but their numbers are controlled. Surveys have shown that the iguanas released on Baltra have started to breed. Capturing wild iguanas that are carrying eggs, housing them in the breeding center, and incubating the eggs has proved effective. The young are raised for two years to give them a better chance of survival when released. Over 800 young iguanas have been released on various islands since the program started. On Isoltes Venecia iguanas have been released into a semicaptive situation, and breeding levels there surpass those of the center.

Most of the Galápagos Islands are now protected, and some areas are out of bounds to all except scientists. Visitors are also controlled to prevent disturbing animals or spoiling the vegetation.

**The Galápagos land iguana** *is a distinctive-looking tropical lizard. Females lay eggs in burrows; males defend the territory around burrows by head bobbing, posturing, biting, and tail-whipping.*

# Galápagos Marine Iguana

## *Amblyrhynchus cristatus*

*The Galápagos marine iguana is a large tropical South American lizard that inhabits the Galápagos Islands west of Ecuador. It is vulnerable to changes in weather and water conditions caused by the periodic warming of the Pacific Ocean known as the El Niño effect.*

The Galápagos marine iguana inhabits rocky seashores, diving into the sea to feed. Its adaptation to its habitat and its dependence on marine algae for food makes it unique among lizards. Its ancestors are thought to have arrived in the Galápagos Islands by rafting on floating vegetation some 10 million years ago. It is possible that the ancestral forms evolved to give rise to both the marine iguana and the land iguana. There are about seven subspecies of marine iguana living mainly on uninhabited islands in the Galápagos. The population on each island varies slightly, either in color or in shape.

### Behavioral Adaptations

Marine iguanas are ectotherms—they need to achieve a certain body temperature before they can follow their normal daily activities. Once they have warmed up by basking in the sun, they dive into the sea to forage on marine algae. The sea around the islands is cold: To maintain their internal body temperature, blood is drawn into the inner organs and brain to conserve heat, and the heartbeat slows to conserve oxygen. The iguanas must bask again before they become too chilled. Young iguanas forage only on the rocks, flats, and parts exposed at low tide. A large amount of salt is ingested during feeding and expelled by being sprayed through the nostrils—often during squabbles.

Males are territorial during the three-month breeding season and can be aggressive toward humans if molested: They will hiss and bite rather than give up their basking stations. Females move inland to bury their eggs in loose soil. They guard the nest for a few days before leaving the eggs to incubate. While protecting the nest, they are vulnerable to predators.

### DATA PANEL

**Galápagos marine iguana**

***Amblyrhynchus cristatus***

**Family:** Iguanidae

**World population:** More than 20,000

**Distribution:** Galápagos Islands

**Habitat:** Rocky seashores

**Size:** Length: up to 5.6 ft (1.7 m). Weight: 20 lb (9 kg)

**Form:** Dark-gray to black iguana; color lightens or darkens to reflect or absorb heat. Males sometimes develop reddish spots during courtship. Long tail is flattened to function as a paddle when swimming. Spiny dorsal (back) crest; prominent granular scales on head. Strong limbs and claws provide a firm grip when climbing over rocky cliffs and ledges

**Diet:** Marine algae

**Breeding:** Female moves as far as 984 ft (300 m) inland to bury eggs in loose soil; 1–6 eggs laid

**Related endangered species:** Galápagos land iguana *(Conolophus subcristatus)* VU

**Status:** IUCN VU

### Threats to Survival

Before humans settled on the Galápagos Islands, marine iguanas had only a few natural predators. Juveniles were at risk from certain birds, crabs, and fish, for example. However, when settlers arrived, they brought dogs and cats, and predation increased. Fortunately, people themselves did not pose a threat. The iguana's diet of marine algae did not bring it into competition with people. Moreover, the islanders did not eat the iguanas or their eggs, nor were they interested in their skins.

**The Galápagos marine iguana** *has strong claws to help it cling to rocks and ledges. When swimming, its long, flattened tail acts as an effective paddle.*

The Galápagos Islands are well protected because of the density of unique species there. The marine iguana benefits from such protection. There are visitor centers on several islands, but many areas are off limits to tourists. (However, scientists or film crews wanting to study Galápagos wildlife can apply for permission to visit the islands.) Collection for the pet trade has never been a problem; creating a suitable habitat in captivity would be impossible, which is why the species does not appear in zoos.

The marine iguana's IUCN listing as Vulnerable is based on the vulnerability of its limited habitat, its susceptibility to introduced predators, and a recent drop in numbers. Perhaps the most significant threat to the Galápagos marine iguana's survival is the El Niño phenomenon. El Niño is an ocean current that usually brings cold, nutrient-rich water to the Galápagos. Periodically, the current is diluted and warmer; low-saline, nutrient-poor water replaces the normal current. Strong El Niño events occurred in 1982 and again in 1997, with several weaker ones, including in 2009. Although plants and animals benefited from the increased rainfall, the change in water quality affected marine creatures, including the marine iguana. A drastic reduction in green marine algae and an increase in inedible, possibly poisonous, brown algae had a severe effect on iguana numbers. On one island alone the marine iguana population fell by an estimated 90 percent after the 1997 El Niño, but numbers are still recovering. An oil spillage in January 2001 killed 60 percent of the marine iguanas on the island of Santa Fé. However, the total population in the archipelago is now believed to be more than 20,000.

EX
EW
CR
EN
VU
NT
LC
O

# Totoaba

## *Totoaba macdonaldi*

*It seems odd that demand for a particular soup could have been a major factor in bringing the totoaba fish to the edge of extinction. The soup—made from totoaba swim bladder—is a culinary delight in Southeast Asia, thousands of miles from the totoaba's homeland.*

The totoaba was once numerous in the Gulf of California. According to some reports, it could be found from the mouth of the Colorado River southward to Mulegé on the western side of the gulf, and Mazatlán on the eastern side. Other reports placed the western range as far down as Bahia Concepción and the eastern one to the Rio El Fuerte around Los Mochis. Today the distribution is much reduced and confined to the northern part of the Gulf of California (Sea of Cortéz).

During the 1920s the totoaba—the largest of the "drums" (bony fish that make a drumming noise)—was fished almost exclusively for its swim bladder. The organ allows fish to control their buoyancy; air can be pumped into or extracted from the swim bladder, allowing the fish to float or sink. The totoaba's swim bladder was dried (it was referred to as "seen kow" when in this state) and then cooked and eaten or used as stock for soup. Totoaba was considered a delicacy by members of the Chinese community in Guaymas, a small town on the mainland coast of the Gulf of California. Swim-bladder soup was already well known to the Chinese inhabitants, so finding a rich source of the delicacy on their doorstep in the form of 3-pound (1.5-kg) swim bladders from the totoaba presented exciting possibilities for commerce.

### DATA PANEL

**Totoaba**

***Totoaba macdonaldi (Cynoscion macdonaldi)***

**Family:** Sciaenidae

**World population:** Unknown

**Distribution:** Gulf of California, Mexico; mainly northern parts

**Habitat:** Deep marine waters; shallow estuarine or marshy, brackish (salty) waters during spawning season

**Size:** Length: up to 6.6 ft (2 m). Weight: up to 300 lb (135 kg)

**Form:** Perchlike fish; dull-colored on top, fading to lighter tones below; 2 dorsal (back) fins (1 with a deep notch); posterior edge of first just touches anterior edge of second; large head and mouth; lateral line organ (pressure-sensitive organ) on side from head back into tail

**Diet:** Fish; shrimp and other crustaceans

**Breeding:** Migrates in spring to shallow estuarine or marshy brackish areas in delta of Colorado River for spawning. After spawning, adults migrate south, into deeper water. Juveniles thought to stay in northern Gulf for about 2 years

**Related endangered species:** None

**Status:** IUCN CR

### Bladder Boom ... and Collapse

Gradually, the market grew for dried totoaba bladders, especially after experimental shipments to Southeast Asia were undertaken by Mexican entrepreneurs. As news of the lucrative industry spread, further fishing, drying, and exporting interests moved into the area. Trade was so successful that it soon led to the disappearance of the totoaba from local waters. Fishing operations followed the fish north, ending up at the top end of the Gulf, where the Colorado River enters the sea, and where totoaba were particularly abundant.

The industry boomed, and people flocked to the area, to the extent that three new villages were established. Over the next 20 years or so catches increased as more sophisticated fishing technology was implemented. The

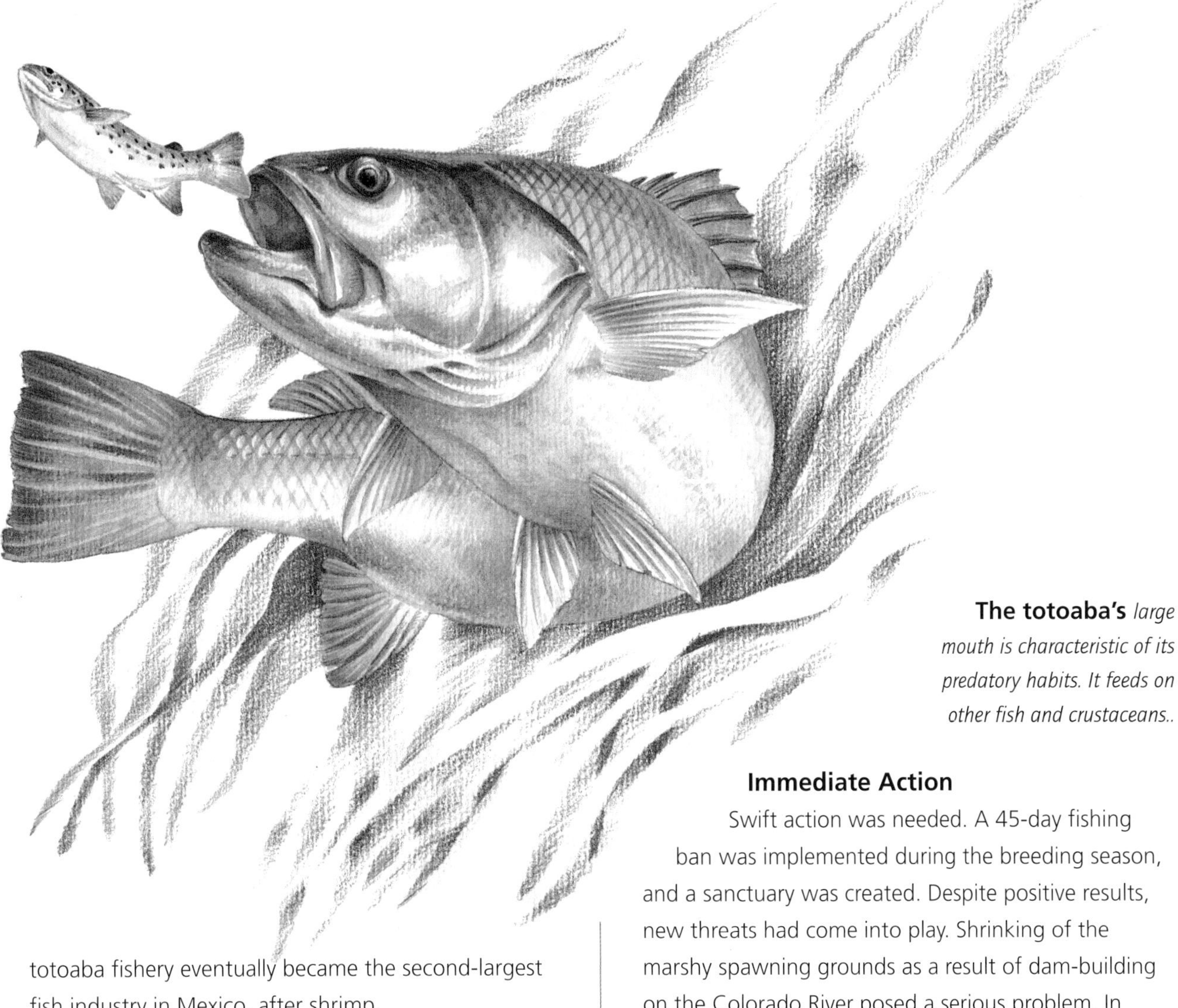

**The totoaba's** *large mouth is characteristic of its predatory habits. It feeds on other fish and crustaceans..*

totoaba fishery eventually became the second-largest fish industry in Mexico, after shrimp.

However, eventually catches began to decline, reaching their lowest in 1958. A combination of factors brought about the collapse of the industry, the sheer numbers of fish collected being only one of them. In addition, large numbers of juveniles were being caught by shrimp trawlers, thus affecting replenishment of stocks. The overriding factor, however, was the fact that the area around the mouth of the Colorado River was where the totoaba migrated to spawn. Mature specimens collected during the breeding season in spring meant a significant decrease in the reproductive capacity of the species as a whole, which, in turn, meant that the juveniles being caught in shrimp nets could not be replaced.

### Immediate Action

Swift action was needed. A 45-day fishing ban was implemented during the breeding season, and a sanctuary was created. Despite positive results, new threats had come into play. Shrinking of the marshy spawning grounds as a result of dam-building on the Colorado River posed a serious problem. In addition, the reduction of freshwater flow into the Gulf of California as a result of the dams led to an increase in salinity in the remaining spawning grounds, exerting yet another negative effect on the totoaba.

By 1979, although commercial fishing had been banned four years earlier, the totoaba was declared an endangered species; some scientists predicted that the species would be extinct by 2000. The establishment of a biosphere reserve in the Upper Gulf in 1993 has been beneficial, and captive breeding may also help protect the totoaba in years to come. However, there is concern in some quarters about the management of the Colorado River. Although it is not yet extinct, there are fears for the fate of the totoaba.

# Categories of Threat

The status categories that appear in the data panel for each species throughout this book are based on those published by the International Union for the Conservation of Nature (IUCN). They provide a useful guide to the current status of the species in the wild, and governments throughout the world use them when assessing conservation priorities and in policy-making. However, they do not provide automatic legal protection for the species.

Animals are placed in the appropriate category after scientific research. More species are being added all the time, and animals can be moved from one category to another as their circumstances change.

## Extinct (EX)

A group of animals is classified as EX when there is no reasonable doubt that the last individual has died.

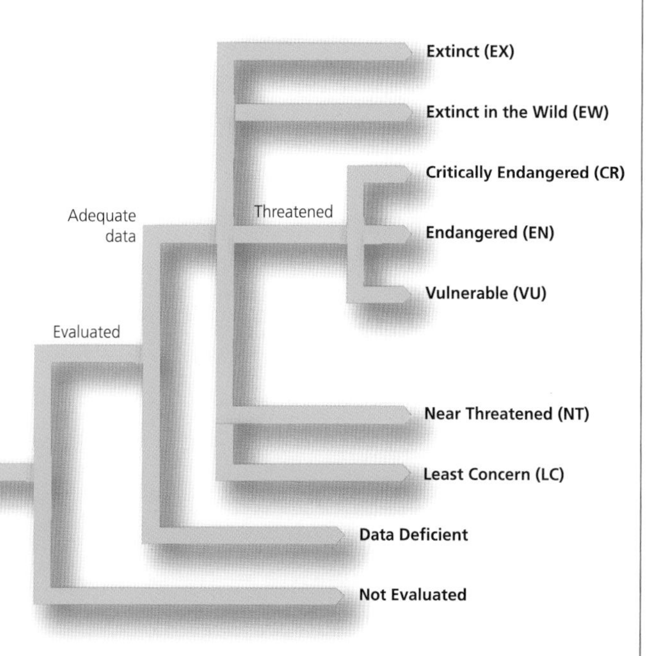


## Extinct in the Wild (EW)

Animals in this category are known to survive only in captivity or as a population established artificially by introduction somewhere well outside its former range. A species is categorized as EW when exhaustive surveys throughout the areas where it used to occur consistently fail to record a single individual. It is important that such searches be carried out over all of the available habitat and during a season or time of day when the animals should be present.

## Critically Endangered (CR)

The category CR includes animals facing an extremely high risk of extinction in the wild in the immediate future. It includes any of the following:

- Any species with fewer than 50 individuals, even if the population is stable.
- Any species with fewer than 250 individuals if the population is declining, badly fragmented, or all in one vulnerable group.
- Animals from larger populations that have declined by 80 percent within 10 years (or are predicted to do so) or three generations, whichever is the longer.

**The IUCN categories** *of threat. The system displayed has operated for new and reviewed assessments since January 2001.*

• Species living in a very small area—defined as under 39 square miles (100 sq. km).

## Endangered (EN)

A species is EN when it is not CR but is nevertheless facing a very high risk of extinction in the wild in the near future. It includes any of the following:

• A species with fewer than 250 individuals remaining, even if the population is stable.

• Any species with fewer than 2,500 individuals if the population is declining, badly fragmented, or all in one vulnerable subpopulation.

• A species whose population is known or expected to decline by 50 percent within 10 years or three generations, whichever is the longer.

• A species whose range is under 1,900 square miles (5,000 sq. km), and whose range, numbers, or population levels are declining, fragmented, or fluctuating wildly.

• Species for which there is a more than 20 percent likelihood of extinction in the next 20 years or five generations, whichever is the longer.

## Vulnerable (VU)

A species is VU when it is not CR or EN but is facing a high risk of extinction in the wild in the medium-term future. It includes any of the following:

• A species with fewer than 1,000 mature individuals remaining, even if the population is stable.

• Any species with fewer than 10,000 individuals if the population is declining, badly fragmented, or all in one vulnerable subpopulation.

**The reason that** *grizzly bears are considered to be of "Least Concern"—rather than a more rare category—is because they live in such a large area of North America and northern Asia. However, this mammal has gone extinct in many other parts of its range, including China, Japan, and Korea.*

• A species whose population is known, believed, or expected to decline by 20 percent within 10 years or three generations, whichever is the longer.
• A species whose range is less than 772 square miles (20,000 sq. km), and whose range, numbers, or population structure are declining, fragmented, or fluctuating wildly.
• Species for which there is a more than 10 percent likelihood of extinction in the next 100 years.

### Near Threatened/Least Concern (since 2001)

In January 2001 the classification of lower-risk species was changed. Near Threatened (NT) and Least Concern (LC) were introduced as separate categories. They replaced the previous Lower Risk (LR) category with its subdivisions of Conservation Dependent (LRcd), Near Threatened (LRnt), and Least Concern (LRlc). From January 2001 all new assessments and reassessments must adopt NT or LC if relevant. But the older categories still apply to some animals until they are reassessed, and will also be found in this book.
• Near Threatened (NT)
Animals that do not qualify for CR, EN, or VU categories now but are close to qualifying or are likely to qualify for a threatened category in the future.
• Least Concern (LC)
Animals that have been evaluated and do not qualify for CR, EN, VU, or NT categories.

### Lower Risk (before 2001)

• Conservation Dependent (LRcd)
Animals whose survival depends on an existing conservation program
• Near Threatened (LRnt)
Animals for which there is no conservation program but that are close to qualifying for VU category.
• Least Concern (LRlc)
Species that are not conservation dependent or near threatened.

**Black-tailed prairie dogs** *are declining in number: while there are still millions of the Arizona species (right), their close neighbors in Utah number only a few thousand.*

### Data Deficient (DD)

A species or population is DD when there is not enough information on abundance and distribution to assess the risk of extinction. In some cases, when the species is thought to live only in a small area, or a considerable period of time has passed since the species was last recorded, it may be placed in a threatened category as a precaution.

### Not Evaluated (NE)

Such animals have not yet been assessed.
**Note: a colored panel** in each entry in this book indicates the current level of threat to the species. The two new categories (NT and LC) and two of the earlier Lower Risk categories (LRcd and LRnt) are included within the band LR; the old LRlc is included along with Data Deficient (DD) and Not Evaluated (NE) under "Other," abbreviated to "O."

**CITES** *lists animals in the major groups in three appendices, depending on the level of threat posed by international trade.*

| | Appendix I | Appendix II | Appendix III |
|---|---|---|---|
| **Mammals** | 297 species<br>23 subspecies<br>2 populations | 492 species<br>5 subspecies<br>5 populations | 44 species<br>10 subspecies |
| **Birds** | 156 species<br>11 subspecies<br>2 populations | 1,275 species<br>2 subspecies | 24 species |
| **Reptiles** | 76 species<br>5 subspecies<br>1 population | 582 species<br>6 populations | 56 species |
| **Amphibians** | 17 species | 113 species | |
| **Fish** | 15 species | 81 species | |
| **Invertebrates** | 64 species<br>5 subspecies | 2,142 species<br>1 subspecies | 17 species<br>3 subspecies |

## CITES APPENDICES

**Appendix I** lists the most endangered of traded species, namely those that are threatened with extinction and will be harmed by continued trade. These species are usually protected in their native countries and can only be imported or exported with a special permit. Permits are required to cover the whole transaction—both exporter and importer must prove that there is a compelling scientific justification for moving the animal from one country to another. This includes transferring animals between zoos for breeding purposes. Permits are only issued when it can be proved that the animal was legally acquired and that the remaining population will not be harmed by the loss.

**Appendix II** includes species that are not currently threatened with extinction, but that could easily become so if trade is not carefully controlled. Some common animals are listed here if they resemble endangered species so closely that criminals could try to sell the rare species pretending they were a similar common one. Permits are required to export such animals.
**Appendix III** species are those that are at risk or protected in at least one country. Other nations may be allowed to trade in animals or products, but they may need to prove that they come from safe populations.

CITES designations are not always the same for every country. In some cases individual countries can apply for special permission to trade in a listed species. For example, they might have a safe population of an animal that is very rare elsewhere. Some African countries periodically apply for permission to export large quantities of elephant tusks that have been in storage for years, or that are the product of a legal cull of elephants. This is controversial because it creates an opportunity for criminals to dispose of black market ivory by passing it off as coming from one of those countries where elephant products are allowed to be exported. If you look up the African elephant, you will see that it is listed as CITES I, II, and III, depending on the country location of the different populations.

# Organizations

The human race is undoubtedly nature's worst enemy, but we can also help limit the damage caused by the rapid increase in our numbers and activities. There have always been people eager to protect the world's beautiful places and to preserve its most special animals, but it is only quite recently that the conservation message has begun to have a real effect on everyday life, government policy, industry, and agriculture.

Early conservationists were concerned with preserving nature for the benefit of people. They acted with an instinctive sense of what was good for nature and people, arguing for the preservation of wilderness and animals in the same way as others argued for the conservation of historic buildings or gardens. The study of ecology and environmental science did not really take off until the mid-20th century, and it took a long time for the true scale of our effect in the natural world to become apparent. Today the conservation of wildlife is based on far greater scientific understanding, but the situation has become much more complex and urgent in the face of human development.

By the mid-20th century extinction was becoming an immediate threat. Animals such as the passenger pigeon, quagga, and thylacine had disappeared despite last-minute attempts to save them. More and more species were discovered to be at risk, and species-focused conservation groups began to appear. In the early days there was little that any of these organizations could do but campaign against direct killing. Later they became a kind of conservation emergency service—rushing to the aid of seriously threatened animals in an attempt to save the species. But as time went on, broader environmental issues began to receive the urgent attention they needed. Research showed time and time again that saving species almost always comes down to addressing the problem of habitat loss. The world is short of space, and ensuring that there is enough for all the species is very difficult.

Conservation is not just about animals and plants, nor even the protection of whole ecological systems. Conservation issues are so broad that they touch almost every aspect of our lives, and successful measures often depend on the expertise of biologists, ecologists, economists, diplomats, lawyers, social scientists, and businesspeople. Conservation is all about cooperation and teamwork. Often it is also about helping people benefit from taking care of their wildlife. The organizations involved vary from small groups of a few dozen enthusiasts in local communities to vast, multinational operations.

## The IUCN

With so much activity based in different countries, it is important to have a worldwide overview, some way of coordinating what goes on in different parts of the planet. That is the role of the International Union for the Conservation of Nature (IUCN), also referred to as the World Conservation Union. It began life as the International Union for the Preservation of Nature in 1948, becoming the IUCN in 1956. It is relatively new compared to the Sierra Club, Flora and Fauna International, and the Royal Society for the Protection of Birds. It was remarkable in that its founder members included governments, government agencies, and nongovernmental organizations. In the years following the appalling destruction of World War II, the IUCN was born out of a desire to draw a line under the horrors of the past and to act together to safeguard the future.

The mission of the IUCN is to influence, encourage, and assist societies throughout the world to conserve the diversity of nature and natural systems. It seeks

**Grand Teton National Park** *in Wyoming straddles part of the Rocky Mountains and is home to 61 species of mammals, including the American bison, which is Near Threatened.*

to ensure that the use of natural resources is fair and ecologically sustainable. Based in Switzerland, the IUCN has over 1,000 permanent staff and the help of 11,000 volunteer experts from about 180 countries. The work of the IUCN is split into six commissions, which deal with protected areas, policy-making, ecosystem management, education, environmental law, and species survival. The Species Survival Commission (SSC) has almost 7,000 members, all experts in the study of plants and animals. Within the SSC there are Specialist Groups concerned with the conservation of different types of animals, from cats to flamingos, deer, ducks, bats, and crocodiles. Some particularly well-studied animals, such as the African elephant and the polar bear, have their own specialist groups.

Perhaps the best-known role of the IUCN SSC is in the production of the Red Data Books, or Red Lists. First published in 1966, the books were designed to be easily updated, with details of each species on a different page that could be removed and replaced as new information came to light. By 2010 the Red Lists include information on about 45,000 types of animal, of which almost 10,000 are threatened with extinction. Gathering this amount of information together is a huge task, but it provides an invaluable conservation resource. The Red Lists are continually updated and are now available online. The Red Lists are the basis for the categories of threat used in this book.

### CITES

CITES is the Convention on International Trade in Endangered Species of Wild Fauna and Flora (also known as the Washington Convention, since it first came into force after an international meeting in Washington D.C. in 1973). Currently 175 nations have agreed to implement the CITES regulations. Exceptions to the convention include Iraq and North Korea, which, for the time being at least, have few trading links with the rest of the world. Trading in animals and their body parts has been a major factor in the decline of some of the world's rarest species. The IUCN categories draw attention to the status of rare species, but they do not confer any legal protection. That is done through national laws.

Conventions serve as international laws. In the case of CITES, lists (called appendices) are agreed on internationally and reviewed every few years. The appendices list the species that are threatened by international trade. Animals are assigned to Appendix I when all trade is forbidden. Any specimens of these species, alive or dead (or skins, feathers, etc.), will be confiscated by customs at international borders, seaports, or airports. Appendix II species can be traded internationally, but only under strict controls. Wildlife trade is often valuable in the rural economy, and this raises difficult questions about the relative importance of animals and people. Nevertheless, traders who ignore CITES rules risk heavy fines or imprisonment. Some rare species—even those with the highest IUCN categories (many bats and frogs, for example)—may have no CITES protection simply because they have no commercial value. Trade is then not really a threat.

## WILDLIFE CONSERVATION ORGANIZATIONS

**BirdLife International**
BirdLife International is a partnership of 60 organizations in more than 100 countries. Most partners are national conservation groups such as the Canadian Nature Federation. Others include large bird charities such as the Royal Society for the Protection of Birds in Britain. By working together within BirdLife International, even small organizations can be effective globally.
www.birdlife.org

**Conservation International (CI)**
Founded in 1987, Conservation International works closely with the IUCN and has a similar multinational approach.
www.conservation.org

**Durrell Wildlife Conservation Trust (DWCT)**
The Durrell Wildlife Conservation Trust was founded by the British naturalist and author Gerald Durrell in 1963. The trust is based at Durrell's world-famous zoo on Jersey in the Channel Islands and has helped many species from extinction, including the pink pigeon, Mauritius kestrel, Waldrapp ibis, and St. Lucia parrot.
www.durrell.org

**Fauna & Flora International (FFI)**
Founded in 1903, this organization has had various name changes. It began life as a society for protecting large mammals, but has broadened its scope. It was involved in saving the Arabian oryx from extinction.
www.fauna-flora.org

**National Audubon Society**
John James Audubon was an American naturalist and wildlife artist who died in 1851, 35 years before the society that bears his name was founded. The first Audubon Society was established by George Bird Grinnell in protest against the appalling overkill of birds for meat, feathers, and sport. By the end of the 19th century there were Audubon Societies in 15 states, and they later became part of the National Audubon Society, which funds scientific research programs, publishes magazines and journals, manages wildlife sanctuaries, and advises state and federal governments on conservation issues.
www.audubon.org

**The Sierra Club**
The Sierra Club was started in 1892 by John Muir. It was through his efforts that the first national parks, including Yosemite, Sequoia, and Mount Rainier, were established. Today the Sierra Club remains dedicated to the preservation of wild places for the benefit of wildlife and of people.
www.sierraclub.org

**World Wide Fund for Nature (WWF)**
The World Wide Fund for Nature, formerly the World Wildlife Fund, was born in 1961. It was a joint venture between the IUCN, several existing conservation organizations, and a number of successful businesspeople. WWF was big, well-funded, and high profile from the beginning. Its familiar giant panda emblem is instantly recognizable.
www.wwf.org

# More Endangered Animals

*This is the second series of* Facts at Your Fingertips: Endangered Animals. *Many other endangered animals were included in the first series, which was broken down by animal class, as follows:*

**BIRDS**

Northern Brown Kiwi
Galápagos Penguin
Bermuda Petrel
Andean Flamingo
Northern Bald Ibis
White-headed Duck
Nene
Philippine Eagle
Spanish Imperial Eagle
Red Kite
California Condor
Mauritius Kestrel
Whooping Crane
Takahe
Kakapo
Hyacinth Macaw
Pink Pigeon
Spotted Owl
Bee Hummingbird
Regent Honeyeater
Blue Bird of Paradise
Raso Lark
Gouldian Finch

**FISH**

Coelacanth
Great White Shark
Common Sturgeon
Danube Salmon
Lake Victoria Haplochromine Cichlids
Dragon Fish
Silver Shark
Whale Shark
Northern Bluefin Tuna
Masked Angelfish
Big Scale Archerfish
Bandula Barb
Mekong Giant Catfish
Alabama Cavefish
Blind Cave Characin
Atlantic Cod
Mountain Blackside Dace
Lesser Spiny Eel
Australian Lungfish
Paddlefish
Ornate Paradisefish
Knysna Seahorse
Spring Pygmy Sunfish

**INVERTEBRATES**

Broad Sea Fan
Giant Gippsland Earthworm
Edible Sea-Urchin
Velvet Worms
Southern Damselfly
Orange-spotted Emerald
Red-kneed Tarantula
Kauai Cave Wolf Spider
Great Raft Spider
European Red Wood Ant
Hermit Beetle
Blue Ground Beetle
Birdwing Butterfly
Apollo Butterfly
Avalon Hairstreak Butterfly
Hermes Copper Butterfly
Giant Clam
California Bay Pea Crab
Horseshoe Crab
Cushion Star
Freshwater Mussel
Starlet Sea Anemone
Partula Snails

**MAMMALS OF THE NORTHERN HEMISPHERE**

Asiatic Lion
Tiger
Clouded Leopard
Iberian Lynx
Florida Panther
Wildcat
Gray Wolf
Swift Fox
Polar Bear
Giant Panda
European Mink
Pine Marten
Black-footed Ferret
Wolverine
Sea Otter
Steller's Sea Lion
Mediterranean Monk Seal
Florida Manatee
Przewalski's Wild Horse
American Bison
Arabian Oryx
Wild Yak
Ryukyu Flying Fox

**MAMMALS OF THE SOUTHERN HEMISPHERE**

Cheetah
Leopard
Jaguar
Spectacled Bear
Giant Otter
Amazon River Dolphin
Sperm Whale
Blue Whale
Humpback Whale
Proboscis Monkey
Chimpanzee
Mountain Gorilla
Orang-Utan
Ruffed Lemur
African Elephant
Black Rhinoceros
Giant Otter Shrew
Mulgara
Kangaroo Island Dunnart
Marsupial Mole
Koala
Long-beaked Echidna
Platypus

**REPTILES AND AMPHIBIANS**

Blunt-nosed Leopard Lizard
Pygmy Blue-tongued Skink
Komodo Dragon
Hawksbill Turtle
Yellow-blotched Sawback Map Turtle
Galápagos Giant Tortoise
Jamaican Boa
Woma Python
Milos Viper
Chinese Alligator
American Crocodile
Gharial
Gila Monster
Japanese Giant Salamander
Olm
Mallorcan Midwife Toad
Golden Toad
Western Toad
Golden Mantella
Tomato Frog
Gastric-brooding Frog

# GLOSSARY

**adaptation** Features of an animal that adjust it to its environment; may be produced by evolution—e.g., camouflage coloration

**adaptive radiation** Where a group of closely related animals (e.g., members of a family) have evolved differences from each other so that they can survive in different niches

**anterior** The front part of an animal

**arboreal** Living in trees

**bill** The jaws of a bird, consisting of two bony mandibles, upper and lower, and their horny sheaths

**biodiversity** The variety of species and the variation within them

**biome** A major world landscape characterized by having similar plants and animals living in it, e.g., desert, rain forest, forest

**blowhole** The nostril opening on the head of a whale through which it breathes

**breeding season** The entire cycle of reproductive activity, from courtship, pair formation (and often establishment of territory) through nesting to independence of young

**brood** The young hatching from a single clutch of eggs

**canine tooth** A sharp stabbing tooth usually longer than the rest

**carapace** The upper part of a shell in a chelonian

**carnivore** An animal that eats other animals

**carrion** Rotting flesh of dead animals

**chelonian** Any reptile of the order Chelonia, including the tortoises and turtles, in which most of the body is enclosed in a bony capsule

**cloaca** Cavity in the pelvic region into which the alimentary canal, genital, and urinary ducts open

**diurnal** Active during the day

**DNA** (deoxyribonucleic acid) The substance that makes up the main part of the chromosomes of all living things; contains the genetic code that is handed down from generation to generation

**dormancy** A state in which—as a result of hormone action—growth is suspended and metabolic activity is reduced to a minimum

**dorsal** Relating to the back or spinal part of the body; usually the upper surface

**ecology** The study of plants and animals in relation to one another and to their surroundings

**ecosystem** A whole system in which plants, animals, and their environment interact

**ectotherm** Animal that relies on external heat sources to raise body temperature; also known as "cold-blooded"

**edentate** Toothless; also any animals of the order Edentata, which includes anteaters, sloths, and armadillos

**endemic** Found only in one geographical area, nowhere else

**eutrophication** an increase in the nutrient chemicals (nitrate, phosphate, etc.) in water, sometimes occurring naturally and sometimes caused by human activities, e.g., by the release of sewage or agricultural fertilizers

**extinction** Process of dying out at the end of which the very last individual dies, and the species is lost forever

**feral** Domestic animals that have gone wild and live independently of people

**fluke** Either of the two lobes of the tail of a whale or related animal; also a type of flatworm, usually parasitic

**gene** The basic unit of heredity, enabling one generation to pass on characteristics to its offspring

**gestation** The period of pregnancy in mammals, between fertilization of the egg and birth of the baby

**harem** A group of females living in the same territory and consorting with a single male

**herbivore** An animal that eats plants (grazers and browsers are herbivores)

**hibernation** Becoming inactive in winter, with lowered body temperature to save energy. Hibernation takes place in a special nest or den called a hibernaculum

**homeotherm** An animal that can maintain a high and constant body temperature by means of internal processes; also called "warm-blooded"

**inbreeding** Breeding among closely related animals (e.g., cousins), leading to weakened genetic composition and reduced survival rates

**incubation** The act of keeping eggs warm for the period from laying the eggs to hatching

**insectivore** Animal that feeds on insects. Also used as a group name for hedgehogs, shrews, moles, etc.

**keratin** Tough, fibrous material that forms hair, feathers, nails, and protective plates on the skin of vertebrate animals

**mammal** Any animal of the class Mammalia—a warm-blooded vertebrate having mammary glands in the female that produce milk with which it nurses its young. The class includes bats, primates, rodents, and whales

**metabolic rate** The rate at which chemical activities occur within animals, including the exchange of gasses in respiration and the liberation of energy from food

**omnivore** An animal that eats a wide range of both animal and vegetable food

**parasite** An animal or plant that lives on or within the body of another (the host) from which it obtains nourishment. The host is often harmed by the association

**pheromone** Scent produced by animals to enable others to find and recognize them

**placenta** The structure that links an embryo to its mother during pregnancy, allowing exchange of chemicals between them

**plastron** The lower shell of chelonians

**posterior** The hind end or behind another structure

**quadruped** Any animal that walks on four legs

**raptor** Bird with hooked bill and strong feet with sharp claws (talons) for seizing, killing, and dealing with prey; also known as birds of prey

**reptile** Any member of the class of cold-blooded vertebrates, Reptilia, including crocodiles, lizards, snakes, tortoises, turtles, and tuataras. Reptiles are characterized by an external covering of scales or horny plates. Most are egg-layers, but some give birth to live young

**scute** Horny plate covering live tissue beneath

**swim bladder** A gas- or air-filled bladder in fish; by taking in or exhaling air, the fish can alter its buoyancy

**underfur** Fine hairs forming a dense, woolly mass close to the skin and underneath the outer coat of stiff hairs in mammals

**vertebrate** Animal with a backbone (e.g., fish, mammal, reptile), usually with skeleton made of bones, but sometimes softer cartilage

**viviparous** (of most mammals and a few other vertebrates) Giving birth to active young rather than laying eggs

# FURTHER RESEARCH

## Books


### Mammals

Macdonald, David, *The New Encyclopedia of Mammals,* Oxford University Press, Oxford, U.K., 2009

Payne, Roger, *Among Whales*, Bantam Press, U.S., 1996

Reeves, R. R., and Leatherwood, S., *The Sierra Club Handbook of Whales and Dolphins of the World*, Sierra Club, U.S., 1988

Sherrow, Victoria, and Cohen, Sandee, *Endangered Mammals of North America*, Twenty-First Century Books, U.S., 1995

Whitaker, J. O., Audubon Society *Field Guide to North American Mammals,* Alfred A. Knopf, New York, U.S., 1996

Wilson, Don E., Mittermeier, Russell A., *Handbook of Mammals of the World Vol 1,* Lynx Edicions, Barcelona, Spain, 2009

### Birds

Attenborough, David, *The Life of Birds,* BBC Books, London, U.K., 1998

BirdLife International, *State of the World's Birds: Indicators for our Changing World*, BirdLife International, Cambridge, U.K., 2008

del Hoyo, J., Elliott, A., and Sargatal, J., eds, *Handbook of Birds of the World Vols 1 to 15,* Lynx Edicions, Barcelona, Spain, 1992-2013

Dunn, Jon, and Alderfer, Jonathan K., *National Geographic Field Guide to the Birds of North America,* National Geographic Society, Washington D.C., U.S., 2006.

Harris, Tim, *Migration Hotspots of the World*, Bloomsbury/RSPB, London, U.K., 2013.

Stattersfield, A., Crosby, M., Long, A., and Wege, D., eds., *Endemic Bird Areas of the World: Priorities for Biodiversity Conservation,* BirdLife International, Cambridge, U.K., 1998

### Fish

Buttfield, Helen, *The Secret Lives of Fishes*, Abrams, U.S., 2000

Dawes, John, and Campbell, Andrew, eds, *The New Encyclopedia of Aquatic Life, Facts On File*, New York, U.S., 2004

### Reptiles and Amphibians

Corton, Misty, *Leopard and Other South African Tortoises,* Carapace Press, London, U.K., 2000

Ernst, Carl H., and Lovich, Jeffrey E., *Turtles of the United States and Canada*. Johns Hopkins University Press, Baltimore, U.S., 2009.

Hofrichter, Robert, *Amphibians: The World of Frogs, Toads, Salamanders, and Newts*, Firefly Books, Canada, 2000

Stafford, Peter, *Snakes*, Natural History Museum, London, U.K., 2000

Taylor, Barbara, and O'Shea, Mark, *Great Big Book of Snakes and Reptiles*, Hermes House, London, U.K., 2006.

### Insects

Eaton, Eric R. and Kaufman, Kenn, *Kaufman Field Guide to Insects of North America*, Houghton Mifflin, New York, U.S., 2007

Brock, Jim P., and Kaufman, Kenn. *Kaufman Field Guide to Butterflies of North America*, Houghton Mifflin, New York, U.S., 2006

### General

Allaby, Michael, *A Dictionary of Ecology*, Oxford University Press, New York, U.S., 2010

Douglas, Dougal, and others, *Atlas of Life on Earth*, Barnes & Noble, New York, U.S., 2001


## Websites

*www.nature.nps.gov* United States National Park Service wildlife site

*www.abcbirds.org* American Bird Conservancy. Articles, information about bird conservation in the Americas

*www.birdlife.org* The site of BirdLife International, highlighting projects to protect the populations of endangered species

*www.cites.org* CITES and IUCN listings. Search for animals by order, family, genus, species, or common name. Location by country and explanation of reasons for listings

*www.cmc-ocean.org* Facts, figures, and quizzes about marine life

*www.darwinfoundation.org* Charles Darwin Foundation

*www.fauna-flora.org* Information about animals and plants around the world on the site of Flora & Fauna International

*www.endangeredspecie.com*
Information, links, books, and publications about rare and endangered species. Also includes information about conservation efforts and organizations

*forests.org* Includes forest conservation answers to queries

*www.iucn.org* Details of species, IUCN listings, and IUCN publications. Link to online Red Lists of threatened species at: www.iucnredlist.org

*www.wwf.org* World Wide Fund for Nature (WWF). Newsroom, press releases, government reports, campaigns. Themed photogallery

*www.wcs.org* Wildlife Conservation Society. Information on projects to help endangered animals in every continent.

*us.whales.org* Whale and Dolphin Conservation. News, projects, and campaigns. Sightings database

# INDEX

Page numbers in **bold** indicate main references to the animals.